A MAGNETIC-LIKE COMPONENT OF THE SOLAR GRAVITATIONAL FIELD

DARK ENERGY

2nd Edition, Revised

By

Thomas W. Hill, Ph.D.
Physics

Printed in the United States of America by CreateSpace, an Amazon.com company.

Notes regarding this revised edition:

This revised edition reorganizes some materials presented in the previous edition, especially the appendices, and uses a different font to facilitate reading. It also corrects errors and includes a theory applicable to cosmic-level phenomena. We offer a non-relativist approach for the *big bang* that correctly models observed Doppler shifts and affirmatively shows that universe expansion is slowing down under the influence of gravitational attraction. Whether or not the expansion stops and collapse begins will be determined by a small enegy term that relates the kinetic and gravitational energies. We further show that proper modeling of the energy state of the Milky Way and of its mass estimated by the Gaia mission disproves claims of a significant influence of *dark matter* on the galaxy. The alignment at the galaxy-level of stars possessing magnetic-like fields is identified as a major source of peculiar velocities in the Hubble diagrams and of supposed *dark energy*.

ABSTRACT

The development of gravitational field theory is advanced by including a second form of gravitational force, comparable to magnetism. Modeling a vector potential $\underline{\mathbf{A}}_g$ provides a magnetic-like flux $\underline{\mathbf{B}}_g$ as the curl of $\underline{\mathbf{A}}_g$, *i.e.*, $\underline{\mathbf{B}}_g = \nabla \times \underline{\mathbf{A}}_g$, which is produced by steady state mass currents within stars such as the sun. The effects of $\underline{\mathbf{A}}_g$ and $\underline{\mathbf{B}}_g$ are evident in the planetary orbits, with $\underline{\mathbf{A}}_g$ setting their inclinations and $\underline{\mathbf{B}}_g$ causing principal frame rotations.

Success of the solar model leads to the hypothesis that about 5 billion years ago, when the sun formed, myriads of stars generated magnetic-like forces and clustered to form galaxy-level phenomena. This approach offers defensible explanations for peculiar velocities, the onset of a second factor in the Hubble law, and supposed *dark energy*. The vector potential $\underline{\mathbf{A}}_{gm}$ is responsible for the development of spiral galaxies, and $\underline{\mathbf{B}}_{gm}$ produces non-central forces that are responsible for clustering at rates much faster than would be achieved under Newtonian forces alone.

Metaphysical notions are rejected, and the reality of gravitational field propagation by waves is emphasized. The nth orbital state in the solar field is based on constant orbital energy E_n and can be derived from a Schroedinger type of wave equation without particle statistics. A specific angular momentum constant σ for the field affords a Fourier relationship between planet position and velocity, where $\mu_p\sigma$ for a planet mass μ_p is comparable to the reduced Planck constant $\hbar$ of atomic theory. A term $2E_n/(\mu_p\sigma^2)$ separates the equation's spatial $\Psi(x)$ component from its temporal $\Im(t)$ component. $\Im(t)$ is the solution of $d^2\Im/dt^2 + [2E_nc^2/(\mu\sigma^2)]\,\Im = 0$, where c is the wave velocity and $\omega_n^{\,2} = 2E_nc^2/(\mu_p\sigma^2)$ is the perfect square required for oscillatory wave behavior as a function of time.

By solving an expanded form of $\nabla^2\Psi + [2E_n/(\mu\sigma^2)]\,\Psi = 0$, where ∇ is the gradient operator, state stability is obtained at specified distances from the sun. However, orbit populations depend on the availability of mass at the time of planet formations and not all allowed states are occupied. The unperturbed results for planetary orbit inclinations and their mean radii agree with observations to the third significant digit. The existence of $\underline{\mathbf{A}}_g$, $\underline{\mathbf{B}}_g$, and a Fourier relationship between planet position and velocity has long been evident, but bias in the physics community has prevented their recognition. †

Body precessions for the Earth and Mars are also modeled by using orbit-level reference frames based on the presence of $\underline{\mathbf{A}}_g$ and $\underline{\mathbf{B}}_g$. The Earth's Chandler Wobble and its far-term nutation are correctly derived, using as inputs the Earth's observed oblateness and southward migration of the Tropic of Cancer. The nutation model provides an average period of about 105,000 years for Earth's Ice Ages, in agreement with the Milanković theory and research results. Estimates for Mars are inconclusive because of lack of data.

The thesis further addresses the perihelion advance of Mercury's orbit, and anomalies observed in the trajectories of Pioneer 10 and 11 spacecraft. Plausible sources for the advance are examined, including general relativity, a gravitational form of Larmor precession, and a quadrupole moment in the solar horizontal plane. The application of $\underline{\mathbf{B}}_g$ to the Pioneer trajectories serves to explain their observed anomalies and gradual extinctions.

A comparison of gravitational theory to electromagnetism is provided under indications that gravity waves propagate at the speed of light, and the equivalents of Maxwell's equations are applied for gravity waves. Also included are proper modeling of phase invariance evidenced by the Michelson-Morley experiments and of velocities attendant to Doppler shifts, which have been misinterpreted by the special theory of relativity. Neither the results of the Michelson-Morley experiments nor observations of starlight aberration depend on the special theory for explanation, but on correctly modeling the movement of detector position as a ray propagates at the constant velocity c.

The effort ends with a non-relativistic cosmology theory that derives Hubble's law following a *big bang* and provides a straightforward model for the expansion of the universe. It is affirmatively shown that the expansion is slowing under gravitational influences. In addition, proper modeling of the Milky Way's star distribution in cylindrical coordinates refutes claims of large amounts of dark matter in its mass, and the spiral appearance is explained by including a vector potential in the galaxy's state equation.

---

† Publication of principal findings of this work by *Physical Review D* was rejected by Prof. Eric Weinberg, Columbia University, under his assessment that the effort is based on "ad hoc assumptions" and "presents nothing new". Cf., Shakespeare, Wm., *Romeo and Juliet*, Act II, scene 2, line 1, "He jests at scars that never felt a wound."

FOREWORD

Many of the fantastic claims authoritatively made by theorists regarding the universe have no basis in reality. The claim that *black holes* exist in the cosmos is only a theory. Neither are there worm holes in a *fabric of space-time*. The latter is actually a resurrected form of an *aether*, whose existence was disproved more than a century ago by the Michelson-Morley † and other experiments. The nonexistence of black holes was recently admitted by one of its major proponents, †† whose position has become a pariah for general relativists. Acceptance of the truth would have been inevitable, however, once the existence of a second form of force in the gravitational field had been recognized. The present deficiency in gravitation theory is comparable to formulating electromagnetic theory without including a magnetic field.

Development of gravitation theory has in fact stagnated during the past century due to a religious form of deference by the physics community to the theory of relativity. Counterintuitive claims based on non-verifiable properties of nonlinear, four-dimensional geometry have been cavalierly applied to the universe. Two of the theory's three traditional tests are not independent, and observations of light ray bending by the sun indicate a dependency on viewing geometry, in violation of the theory's equivalence principle.

But the worst failing of the general theory is that major revisions are required whenever new discoveries are made. One example is *inflation* theory, which is actually a second *big bang* wherein the conservation of energy and momentum are casually violated. ††† Even a novice should recognize a logic disconnect in supposing that a single big bang created the entire universe, then arguing for its repetition. Despite being hailed as the proper model for

---

† Michelson and Morley, *Silliman J.*, **34**, 333, 427 (1887). The experiment showed conclusively that the phase of light rays is not affected by the motion of an observer.

†† See, http://www.nature.com/news/stephen-hawking-there-are-no-black-holes-1.14583, stating with some courage, "black holes as we perceive them do not exist".

††† See, *e.g.*, Marmet, P, "The Cosmological Constant and the Redshift of Quasars", pp. 3-5, 11/26/2010, http://www.newtonphysics.on.ca/quasars/index.html.

which Newtonian physics is only an approximation, the theory sheds no light on major unexplained solar system observations, such as the distinct inclinations and spacing of the planetary orbits. Neither does it explain recently observed anomalies in the trajectories of interplanetary spacecraft.

The problem we face is not a flaw in the mathematics of the proposed theories, but rather the correspondence of these theories with reality. The proponents start with a set of equations based primarily on the general theory of relativity and, with adjustments, develop and confidently apply the results to a cosmos we only marginally grasp. These positions then become dogma.

One reality check is provided by reconsideration of the advance of the perihelion of the highly elliptical orbit of the planet Mercury. During every orbit cycle the perihelion – its point of closest approach to the sun – advances by a very small amount in the direction of the orbital motion. Most of this phenomenon is due to tugs on the orbit produced by the gravitational fields of other planets, including the Earth, but a small amount remains unaccounted for. Albert Einstein used the anomaly to promote his general theory, claiming that its source is a curvature of space produced by the sun's large mass, which he modeled by proposing an arbitrary *Schwarzschild radius* as a perturbation to the orbit. But the theory has not been verified for other planetary orbits and does not apply to Earth satellites wherein the same phenomenon is observed and attributed to the Earth's oblateness. Einstein's Schwarzschild radius for the sun is actually twice the gravitational equivalent of the magnetic permeability of free space multiplied by the sun's mass.

Years ago, while America was still a set of British colonies, the brilliant Swiss mathematician and physicist Leonhard Euler (1707-1783) provided a methodology for understanding perihelion advance and analogous phenomena. Briefly stated, torques on an orbit cause a radial-based coordinate system to rotate. The process is appropriately called *frame rotation*, and it adds angular momentum to the orbital component. We use frame rotations below to compare the Einstein correction with two candidate torques for Mercury's orbit – a perturbation in the sun's field due to a small quadrupole moment, and the gravitational equivalent of Larmor precession in a magnetic field.

Prior to addressing the perturbations, we present a standing wave theory for the solar gravitational field in order to explain the observed planetary orbit inclinations and radii. By including orbit-level frame rotations, we provide a previously missing solar field structure for the orbits. The classical methods that we use predate relativity theory and are analogous to those known to apply to electromagnetism. Our theory enables proper modeling of the Earth's spin axis nutation, similar to a spinning top, and provides both the *Chandler Wobble* and spin obliquity limits. The latter result agrees to first order with research data for the average of the Earth's glacial cycles.

Our most important revelations are (1) the existence of a gravitational vector potential $\underline{\mathbf{A}}_g$ due to mass currents in stars such as the sun, and (2) evidence of field propagation via gravity waves. The curl of $\underline{\mathbf{A}}_g$ is a flux $\underline{\mathbf{B}}_g$, which produces a velocity-dependent, non-central force comparable to magnetism and a torque on, and frame rotation for, inclined orbits. The success of our approach in explaining multiple orbit parameters previously regarded as random makes it highly unlikely that the vector potential $\underline{\mathbf{A}}_g$ is fictitious. The expression for $\underline{\mathbf{B}}_g$ modeled for the orbits further explains the onset and magnitude of trajectory anomalies observed for Pioneer 10 and 11 spacecraft. The flux is also applicable to other spacecraft trajectory anomalies for which the heat radiation model adopted for the Pioneers is inappropriate.

The occurrence of $\underline{\mathbf{B}}_g$ in the myriads of stars in galaxies leads to a collective force that impacts Newtonian gravity at relative galaxy velocities and offers an explanation for a phenomenon currently called *dark energy*. Unhappily, concepts of a gravitational vector potential and field propagation by waves are too outrageous for people who argue that the entire universe was once confined in a volume of dimension less than 10^{-33} centimeters.

We have included overviews of the solar system, the Milky Way, and the universe itself, and our results for the above listed subjects are discussed in detail. Readers with limited mathematics skills are encouraged to focus on the narratives of the results. We may envision an eternal universe which reaches a maximum extent over many billions of years, and then collapses under gravitational attraction until a dismembered state occurs. An ensuing explosion spawns billions of galaxies that evolve and expand into limitless space until maximum dispersion is reached, and the process repeats. Observations support this model, which is more plausible than hypotheses that space itself is expanding and the progression of time depends on velocity.

TABLE OF CONTENTS

LIST OF APPENDICES

LIST OF FIGURES AND TABLES

SOME PHYSICAL CONSTANTS

1. Universal gravitational constant:
   $g_o \cong 6.670 \times 10^{-11}$ meter-cubed per kilogram second-squared = $1/(4\pi\varepsilon_g)$,
   where ε_g is the gravitational equivalent of free space permittivity

2. Estimated Masses: Sun: $M_\oplus \cong 1.991 \times 10^{30}$ kilograms, *
   Milky Way: $M_m \cong 4.83 \times 10^{41}$ kilograms,
   Universe: $M_t \cong 5.75 \times 10^{51}$ kilograms,

3. Specific angular momentum constant for the solar field:
   $\sigma \cong 1.946 \times 10^9$ kilometers-squared per second

4. Gravitational fine structure constant: $\alpha_g \cong 1/(4393)$

5. Speed of light: $c \cong 2.99793 \times 10^8$ meters per second

6. Gravitational equivalent of free space permeability:
   $\kappa_g \cong 9.324 \times 10^{-30}$ kilometers per kilogram $\equiv 4\pi g_o/c^2$,
   $M_\oplus\kappa_g/(4\pi) = 1.478$ kilometers = 0.5 Schwarzschild radius

7. Electronic charge: $e \cong 1.6021 \times 10^{-19}$ coulomb

8. Electronic rest mass: $m_e \cong 9.1083 \times 10^{-31}$ kilogram

9. Reduced Planck constant:
   $\hbar = h/(2\pi) \cong 1.05443 \times 10^{-34}$ kilogram meter-squared per second

10. Atomic fine structure constant: $\alpha_e \cong 1/(137.037)$

11. Permittivity of free space:
    $\varepsilon_o \cong 8.8543 \times 10^{-12}$ coulomb-second-squared per kilogram meter-cubed,
    $1/(4\pi\varepsilon_o) \cong 8.9874 \times 10^9$ kgm meter-cubed per coulomb-second- squared

12. One Year $\cong 3.1558 \times 10^7$ seconds

13. MegaParsec (MPc) $\cong 3.086 \times 10^{19}$ kilometers $\cong$ 3.262 million light-years

*$M_\oplus g_o \cong 1.328 \times 10^{11}$ kilometers-cubed per second-squared.

1 – INTRODUCTION

The Origins Of Modern Astronomy

The origins of modern astronomy may be traced to Nicolaus Copernicus (1473-1543 A.D.), a Prussian Pole, who collected first order orbital data for the planets and published a heliocentric theory for the solar system. His efforts were extended by a Dane named Tycho Brahe (1546-1601), who recorded extensive planetary observations, but was unable to relate the data to a mathematical form. One of Brahe's German students, Johannes Kepler (1572-1630), extended the work and formulated the correct expression for the planetary orbits, *i.e.,* "every planet [†] describes an ellipse with the sun at one focus". However, Kepler was unable to derive the result from a physical law, such as a force equation. That task was accomplished by the English physicist Isaac Newton (1642-1727) who, by applying his then newly-developed calculus to the law of gravity, was able to specify the distance r from the nth planet to the sun as $r = a_n (1 - \varepsilon^2)/(1 + \varepsilon \cos\theta)$, where a_n is the orbit's semi-major axis, ε is its eccentricity, and θ is the reference angle between the planet and the sun. See Figure 1.1. Although the form agrees with observation, Newton's solution allows the parameters of the ellipse to be any of an infinite number, so long as $0 \leq \varepsilon \leq 1$. [††]

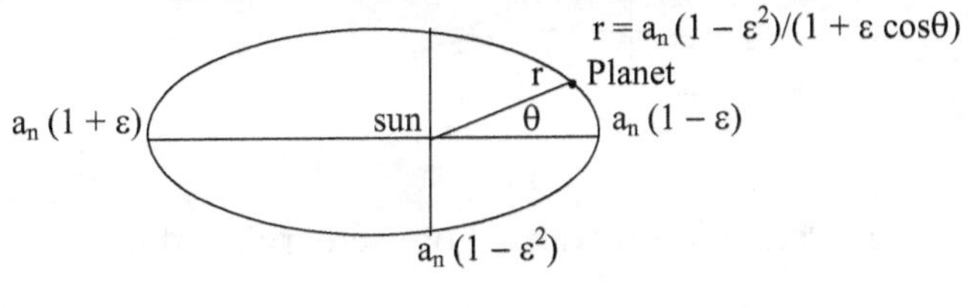

Figure 1.1. Elliptical Orbits For The Planets

---

† The word planet is derived from the Greek verb *planasthal*, meaning to wander, as opposed to the stars which appear to be fixed in the sky. Initially the term included the sun, Moon, and seven planets visible to the naked eye.

†† The form $\varepsilon \leq 1$ indicates that ε is "less than or equal to 1", while $\varepsilon \geq 1$would indicate that ε is "greater than or equal to 1".

In his 1715 *Elements of Astronomy* David Gregory noted a specific ratio between the average value of r for the known planets and the average value of r for the Earth. Later, in a 1766 translation of Charles Bonnet's *Contemplation de la Nature* (1764), Johann D. Titius provided a mathematical formula to describe the relationship. Johann E. Bode publicized the form in his *Anleitung zur Kenntniss des gestirnten Himmels* (1772), which is now called Bode's law, or the Titius-Bode law. Although the "law" exhibits significant errors in the observed planetary orbit values [†] and fails completely for the orbit of Neptune, it has become part of an age old controversy regarding relationships between heavenly body motions, music, [††] and religion, which have persisted since the time of early Greek civilization. Kepler himself advocated a "rhombic solids philosophy", which he alleged as proof that there were exactly six planets in solar orbits -- no more and no less.

The 1500s marked the beginning of astronomy as a true science, characterized by extensive data collection and reduction. Using a crude telescope, Galileo Galilei (1564-1642) observed that the reflected light from Jupiter reveals four moons, which resemble a mini solar system. A total of eight major planets (plus dwarf planets like Pluto), with satellites of various sizes and numerous asteroids in solar orbits have now been catalogued.

perihelion 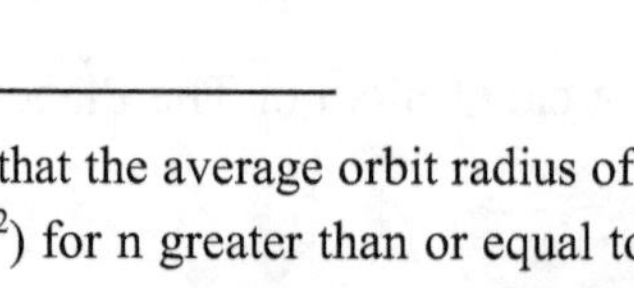

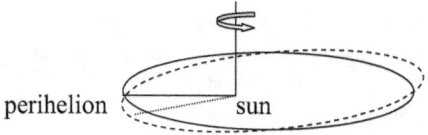

Figure 1.2. Perihelion Advance For Mercury's Orbit (Exaggerated)

[†] Bode's law states that the average orbit radius of nth planet from the sun is given by $a_n = a_1 (2^2 + 3 \times 2^{n-2})$ for n greater than or equal to 2, where $4a_1$ is the orbit radius of Mercury, "×" indicates "multiplied by", and 2^n is 2 multiplied by itself n times.

[††] From the time of the ancient Greeks to the Middle Ages, the motions of the stars were compared to harmonious relationships among musical notes. It was also believed that the heavenly bodies were embedded on spherical surfaces revolving around the Earth. Such perceptions are still reflected in some Christmas songs, for example, which refer to the "music of the spheres".

Dynamic phenomena have also been observed, including a regular advance of the perihelion of Mercury's orbit, depicted in Figure 1.2. Figure 1.1 above provides an illustrative perihelion $a_n(1 - \varepsilon)$ occurring at $\theta = 0$. Another important observation is a general decrease in the frequency of light that originates in the distant stars. This *red shift*, which may be compared to the lower tone of a train whistle as it moves away from us, has been taken to indicate that the universe as a whole is expanding.

Two centuries after Newton, Albert Einstein (1879-1955) proposed a special theory of relativity (which he called *invariance*) to explain the constant group velocity of light waves originating from the distant stars. Using Riemannian geometry, he later sought to develop a general theory to include gravity effects by treating a postulated four-dimensional space-time continuum as being warped by massive objects such as the sun. † After several tries, he successfully modeled that portion of Mercury's perihelion advance not attributable to third body perturbations and, upon inputting an arbitrary cosmological constant, the galactic red shift. He also predicted the bending of light rays as they pass near the sun. Subsequent star observations during the sun's eclipses by the Moon have provided mixed results, raising various explanations for the data obtained at the time of viewing.

It is not unreasonable to regard the general theory as a descriptive, rather than causative, approach which accommodates physical phenomena and models their effects. Since the theory is non linear, the geometry alone produces a change when a vector, such as a force, is moved from one location to another. The theory is difficult to understand, awkward to work with, and almost always requires major approximations of its equations in order to obtain results. Some of its results appear to be nothing more than extraneous mathematical terms not corresponding to any physical phenom-

---

† Of the general theory Nikola Tesla (1856-1943) stated, "[S]pace cannot be curved, for the simple reason that it can have no properties. ... Of properties we can only speak when dealing with matter filling the space. To say that in the presence of large bodies space becomes curved is equivalent to stating that something [matter] can act upon nothing [space] to change the form of nothing [space]."

ena. There is nothing in the theory to constrain the relationships among the planetary orbits, and the Titius-Bode law has remained an unsolved mystery of the universe.

The Planetary Orbits

The planetary orbits are ellipses with the sun located approximately at one focus, as sketched in Figure 1.1 above. † Since the combined system mass is not much greater than the solar mass and the orbit eccentricities are generally small, the ellipses are nearly concentric circles about the sun. Mercury and Pluto are non-circular exceptions, having eccentricities of 0.2056 and 0.2471, respectively. The mean radius of the orbit of the Earth, about 1.496×10^8 kilometers, †† is defined as the astronomical unit (a.u. or AU) and is the baseline used to measure the planetary orbit radii. The smallest mean radius, that of Mercury, is about four-tenths of the Earth's value. Pluto is roughly a hundred times as far from the sun as Mercury.

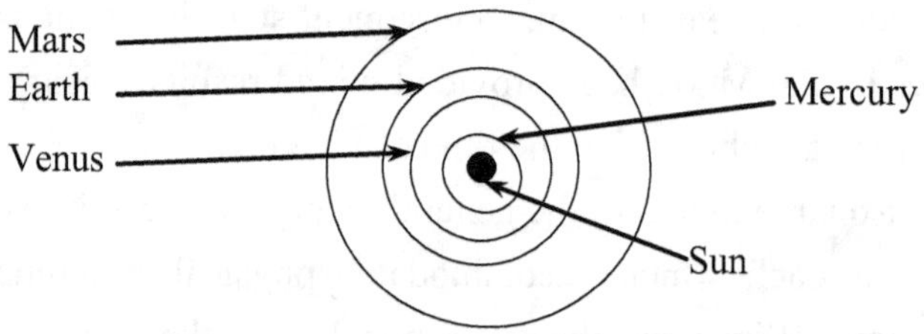

Figure 1.3. Relative Orbit Spacing Of The Terrestrial Planets

---

† A massive attracting source such as the sun actually performs small orbit-like motions in response to the attractive forces of its satellites as they orbit.

†† As indicated above, 10^8 is 10 multiplied by itself eight times, whereas 10^{-8} represents 1 part of 10^8 parts, and $10^{1/2}$ is the square root of 10. The symbol ×, meaning "multiplied by", is often omitted in favor of parentheses or brackets. We may use the latter when one number multiplies another, but forms such as g(x) indicate that g(x) is a function of x. Meters, kilograms, and seconds are standardized measures of distance, mass, and time. A joule is the work required to accelerate a mass of one kilogram at the rate of one meter per second per second over a meter's distance. One joule-second is one joule expended for one second of time.

The first four planets of the solar system are designated as *terrestrial* planets, while the remaining four major planets are called the *outer* planets. Lying in orbits of varying eccentricities between Mars and Jupiter is a belt of separate masses called asteroids, some of which are labeled as minor planets. Figures 1.3 and 1.4 provide rough overviews of the spacing of the planetary orbits. Beyond Pluto's orbit [†] are regions of debris, including icy comets. The first region is the Kuiper Belt and further out is the Oort cloud. Some orbits of comets and asteroids periodically bring them near major planets.

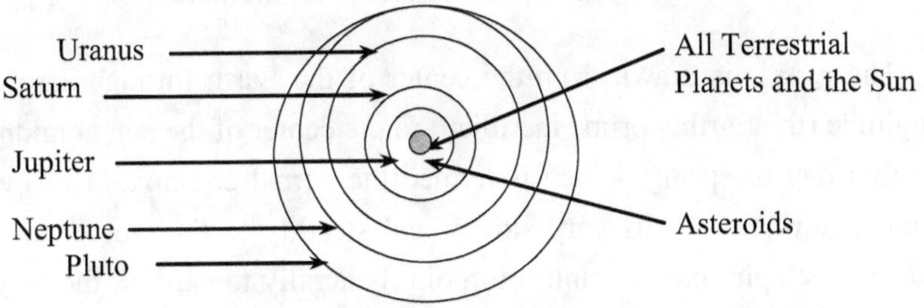

Figure 1.4. Relative Orbit Spacing Of The Outer Planets

The planetary orbits do not lie in the same plane; however, the variations of their inclinations to a common reference axis are not great. Astronomers use the Earth's orbit plane, called the *ecliptic*, as the reference. Figure 1.5 depicts the ecliptic, wherein the $\underline{\mathbf{k}}_e$ axis of an $\underline{\mathbf{i}}_e\underline{\mathbf{j}}_e\underline{\mathbf{k}}_e$ Cartesian coordinate system is perpendicular to the Earth's orbit and defines its orbit vector $\underline{\mathbf{J}}_e$ [††]. The $\underline{\mathbf{j}}_e$ axis is perpendicular to $\underline{\mathbf{k}}_e$ and lies in the plane containing $\underline{\mathbf{J}}_e$ and the Earth's spin vector $\underline{\mathbf{s}}$ indicated in the figure. The $\underline{\mathbf{i}}_e$ axis is perpendicular to both $\underline{\mathbf{j}}_e$ and $\underline{\mathbf{k}}_e$, with the $\underline{\mathbf{i}}_e\underline{\mathbf{j}}_e\underline{\mathbf{k}}_e$ axes forming a right-handed coordinate system.

[†] Pluto has a very small mass and has been re-designated as a "dwarf planet", rather than a major planet. We note that a spaceship leaving the Earth and accelerating constantly at a rate of Earth gravity of 32 feet per second squared would require a year to reach Pluto, at which time it would be approaching the speed of light.

[††] Bold prints with underlined letters represent vectors. Their individual components are indicated in parentheses separated by commas, *e.g.*, $\underline{\mathbf{r}} = (x, y, z)$, or alternatively for unit vectors of $\underline{\mathbf{i}}, \underline{\mathbf{j}},$ and $\underline{\mathbf{k}}$ in a Cartesian coordinate system, $\underline{\mathbf{r}} = x\underline{\mathbf{i}} + y\underline{\mathbf{j}} + z\underline{\mathbf{k}}$. The magnitude of $\underline{\mathbf{r}}$ is $r = |\underline{\mathbf{r}}| = (x^2+y^2+z^2)^{1/2}$.

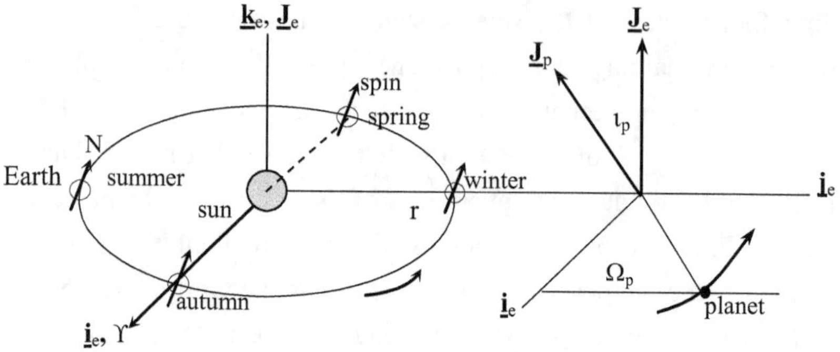

Figure 1.5. Orbit Orientation Parameters

The $\underline{\mathbf{i}}_e$ axis is drawn from the center of the Earth through zero degrees longitude (the Earth's prime meridian) to the center of the sun at midnight on the first day of spring, which is termed the vernal equinox. However, this pointing direction drifts very slowly and steadily to the west from year to year, as the spin axis continues to point generally toward N, the North Star (Polaris). The drift is called the Earth's precession and makes it necessary to specify the direction on a specific date called the epoch. The $\underline{\mathbf{i}}_e$ axis has been given the symbol ϒ, since it was approximately aligned with the first point of the constellation Aries (the Ram) for many years. The $\underline{\mathbf{i}}_e$ axis now points toward the constellation Aquarius, hence the *Age of Aquarius*.

The angle ι_p shown on the right of Figure 1.5 is the inclination of another planet's orbit vector $\underline{\mathbf{J}}_p$ relative to $\underline{\mathbf{J}}_e$. The angle Ω_p designates the angle to the $\underline{\mathbf{i}}_e$ axis at which the planet's orbit intersects the $\underline{\mathbf{i}}_e\underline{\mathbf{j}}_e\underline{\mathbf{k}}_e$ ecliptic plane on its northward journey. All of the planets traverse their orbits in a right-handed, or posigrade, direction. The majority of the planetary spins are also posigrade, excepting Venus, Uranus, and Pluto, whose spins are retrograde (R).

The terrestrial planets have only a few moons -- none for Mercury and Venus, an atypically large Moon for the Earth relative to the parent planet's mass, and two small moons for Mars. Jupiter, Saturn, Uranus, and Neptune have many satellites. Our spacecraft have recently discovered a number of additional small moons and rings for the outer planets. The sun, whose mean density is about 1.4 grams per cubic centimeter, possesses 99.9 per cent of the mass of the solar system. This is about 333,000 times the mass of the Earth. The preponderance of the solar mass is hydrogen; however, it is estimated that the amount of heavy elements in the sun is comparable to the total comprising the planets.

The Need To Improve The Theory Of Gravity

During the period when Einstein was developing his relativity theories, a revolution of thought regarding electromagnetic field interactions with matter was occurring in the study of atomic physics. Classical concepts of infinite divisibility gave way to quantum theories which treated properties of the atom, such as its energy and angular momentum states, as occurring in discrete increments based on the Planck constant, *h*. † Great physicists labored collectively to produce the modern theory of quantum mechanics, which has led to our understanding of the atomic nature of matter. Although Einstein accepted basic quantum theory, as evidenced by his analysis of the photoelectric effect, he refused to embrace the inherent indeterminacy of states, as advocated by many atomic physicists.

The electric field of atomic theory and the gravitational field as propounded by the general theory have proceeded along separate paths, and have not been satisfactorily merged into a unified concept. But the two fields have many common characteristics. Both vary in strength from a point source as $1/r^2$, and in proportion to the product of their measures -- mass for the gravitational field and charge for the electric field. A principal conceptual difference between the two has been lack of consideration of a velocity-dependent force component for gravity. However, we will show below that the observed planetary orbit inclinations and spin obliquities can be explained by the existence of a *gravitational vector potential* $\underline{\mathbf{A}}_g$ in the solar field, with a velocity-dependent flux $\underline{\mathbf{B}}_g$ produced by the curl of $\underline{\mathbf{A}}_g$.

As a non linear field concept, the general theory stands isolated from electromagnetism and other classical theories of physics, although it takes credit for the results of classical gravitational theory, which it treats as an approximation to its more complex modeling. But perhaps we have been too

---

† The energy E contained in an electromagnetic wave of angular frequency ν is given by $E = h\nu$, where the Planck constant *h* is about 6.625×10^{-34} joule seconds.

acquiescent. The general theory is not needed to account for Mercury's 40-plus arc-seconds per century of perihelion advance, which remain after more than 5500 arcseconds due to planetary perturbations have been removed. † By comparison, informed space operations planners know that the advances and retards of perigee for Earth satellites, as well as some other perturbation effects, are caused by a modest asymmetry in the Earth's mass distribution. The quadrupole moment for the Earth, designated as J_2, accounts for the major portion of perigee advance with higher order zonal harmonics (independent of longitude) and tesserals (dependent on longitude and latitude) accounting for the remainder. Figure 1.6 depicts the Earth's quadrupole in the form of a small flattening in its shape and attendant mass at the poles.

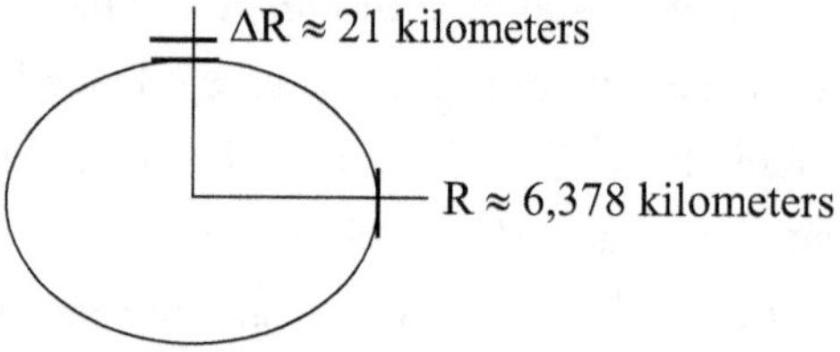

Figure 1.6. Oblateness Of The Earth (Exaggerated)

General theory apologists contend that Earth satellites, whose orbital elements are precisely known, are not good test objects for the theory because of the Earth's "large" oblateness. Upon calculating the perihelion advance for Mercury's orbit, they dismiss the quadrupole model and instead insert a general theory term whose radial dependence is of the same form as the first-order correction for an oblate attracting mass. Visual observations of the sun's outer surface do indicate a spherical shape; however, an overview of the solar system as a whole shows a discoid coalescing with the sun at its center. An equatorial aggregation of massive elements in the sun's interior, corresponding to a J_2 value of about one-sixth of the Earth's known

---

† The importance of the unexplained 43 arcseconds of perihelion advance is evidenced by diverse theories offered by physicists in the early 1900s, including proposed modifications to the inverse square attraction law and claims of the existence of an undiscovered planet.

value, produces the same perihelion advance as the general theory. We also offer below a source for the phenomenon provided by $\underline{\mathbf{B}}_g$, which effect vanishes when the orbit eccentricity is zero. The advance for other planetary orbits should serve to determine which of the theories is correct since the general theory effect is independent of eccentricity and decreases much more slowly with distance than the quadrupole effect. Unfortunately, observations for orbits beyond Mercury have thus far been deemed as inconclusive.

Spacecraft which orbit the sun above its poles should enable us to determine whether the general theory computation, which is independent of elevation angle, is correct. The criticality of this issue cannot be overly emphasized since the perihelion advance is the only one of the classic relativity tests which does not explicitly involve properties of electromagnetic waves. If the general theory should fail fitting tests for precisely-known orbit inclinations, it must be modified. Unhappily, we have not found any report of an attempt to use currently available spacecraft data to differentiate between the possible sources.

A major shortcoming of the theory that need not await further space probe data is an explanation of observed planetary parameters. An objective analyst would suspect that a form of quantization is occurring in the solar field, comparable to stable orbit states modeled by atomic theory. Furthermore, the fact that the terrestrial planetary orbits do not lie in the same horizontal plane, as contrasted with the orbits of the principal moons of Jupiter and Saturn, is an indication that some gravitational effect other than inverse square attraction may be present in the solar field. This suspicion is reinforced by the presence of significant inclined spin for planets other than Mercury and Venus.

Our hypotheses in this regard are that (1) stable orbit states in the field are tilted away from the sun's polar axis as the result of including the gravitational vector potential $\underline{\mathbf{A}}_g$ in the expression of the orbital energy, and (2) $\underline{\mathbf{B}}_g$, the curl of $\underline{\mathbf{A}}_g$, provides a velocity-dependent component of gravitational force $\underline{\mathbf{F}}_B$. The force is negligible below orbital velocities but produces a torque and orbit-level frame rotation whenever a planet's specific orbital angular momentum vector $\underline{\mathbf{J}}_{orb}$ is inclined to the polar axis in an accelerated frame based on planetary spin, as is true for the Earth. We follow atomic electromagnetic theory in our modeling and interpret the methodology to reflect the presence of standing solar gravitational waves that provide regions of stability wherein mutual attraction has enabled diffuse matter to coalesce, forming the planets. Our hypotheses provide a supportable basis for the Titius-Bode law and the non-equatorial, distinct inclinations of the planetary orbits to the polar axis of the sun's gravitational field.

Using the accelerated reference frames provided by this approach, we are able to model the Earth's body motions, including precession of the equinoxes and both near and far term nutations of its spin axis. Our results for the Earth agree with observations. We further show that observed anomalies for Pioneer spacecraft trajectories during long coasting periods beyond Jupiter's orbit can be explained by their interactions with $\underline{\mathbf{B}}_g$.

We additionally dispute the claim that *dark matter* controls spiral galaxy behavior. Proper modeling of the observed exponential mass distributions of these galaxies shows that the tangential velocities of stellar orbits remain almost constant over a wide range of distances from the galaxy centers. This finding, together with observed scale lengths and spiral geometries, validates a wave theory that includes a vector potential $\underline{\mathbf{A}}_g$ at the galaxy level. Acceptance of the existence of the $\underline{\mathbf{B}}_g$ phenomenon in collections of stars also resolves *dark energy* hypotheses posed by general relativists, which are at variance with a straightforward cosmology theory based on the big bang. †

† The "big bang" refers to the form of creation of the universe first proposed by Catholic University of Leuven Professor Georges Lemaître (1894 – 1966). The term was intended to be derisive, and was coined by astronomer Fred Hoyle, who never accepted the concept. The general theory of relativity allows for a big bang, but the concept was initially rejected by Einstein. The theory also allows for black holes in the universe, which are alleged to be huge gravitational sinks. See, Hawking, S., *The Illustrated A Brief History of Time*, Bantam Dell, New York (1988, 1996), Chapters 6, 7. The proponents initially held that nothing could escape from a black hole once it passed an edge called the "event horizon". Hawking subsequently modified the theory to allow for the emission of particles just outside the horizon. However, he recently disavowed claims of the existence black holes, as noted above.

2 – BASIC MATHEMATICAL TOOLS, VECTORS, PLANET PARAMETERS, AND THE LAGRANGIAN

Derivatives And Integrals

Let us briefly review the concept of a derivative in the theory of calculus, † starting with a function represented by g(x) for a variable x. We call x the *independent* variable and g(x) the *dependent* variable, that is dependent on x. For example, x might be the length of the side of a square, and g(x) might be the area of the square. In this example g(x) is simply x multiplied by x, or in algebraic notation, $g(x) = x^2$. The derivative is defined as the instantaneous rate of change of g(x) as x varies. In mathematical terms the derivative is the limit of the difference g(x + dx) – g(x) divided by dx as the small increment dx approaches zero, and is usually written as dg/dx or more compactly as g′(x) in equation 1. The term "limit" is what we expect from the English language, although there are more esoteric ways to state the definition.

$$dg/dx = g'(x) = \text{limit}_{\,dx \to 0}\,[g(x + dx) - g(x)]/dx. \qquad (1)$$

In our example, $g(x+dx) = (x+dx)^2 = x^2 + 2x\,dx + dx^2$. Subtracting g(x) $= x^2$ from the sum, dividing by dx, and setting dx to zero, we find that g′(x) = 2x. Thus, a large value of x leads to a large rate of change in this example. The notion of the derivative may be extended to various types of functions, such as the sine, cosine, and tangent of an angle θ, normally written as sinθ, cosθ, and tanθ, with Greek symbols used to indicate angles. For most functions, we may readily compute the derivatives or look them up in tables.

Figure 2.1 provides a graphic display of the change in an arbitrary function g(x) before dx is set to zero. The value of the derivative at any value of x is the *slope* of the curve at x, and we see that the slope will lie exactly on the curve as dx becomes smaller and smaller. When the independent variable is time, we write it as t rather than x, and using f(t) to express object

---

† Gottfried von Leibnitz (1646-1716) developed the theory of calculus. It was developed independently by Isaac Newton (1642-1727) a few years later.

position as a function of time, $f'(t)$ represents the velocity of the object at any given time. Treating $f'(t)$ as a function of t, we may take its derivative, $d^2f/dt^2 = f''(t)$. This second order derivative is the rate of change of velocity, *i.e.,* acceleration. A force $\underline{\mathbf{F}}$ applied to a mass μ at $\underline{\mathbf{r}} = x\,\underline{\mathbf{i}}_o + y\,\underline{\mathbf{j}}_o + z\,\underline{\mathbf{k}}_o$ causes it to accelerate according to $\underline{\mathbf{F}}/\mu = d^2x/dt^2\,\underline{\mathbf{i}}_o + d^2y/dt^2\,\underline{\mathbf{j}}_o + d^2z/dt^2\,\underline{\mathbf{k}}_o = d^2\underline{\mathbf{r}}/dt^2$, where $\underline{\mathbf{i}}_o$, $\underline{\mathbf{j}}_o$, and $\underline{\mathbf{k}}_o$ are unit vectors in an inertial frame.

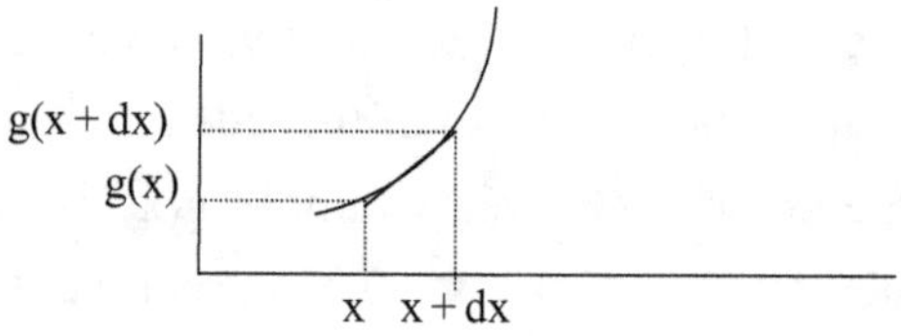

Figure 2.1. Graphic Representation Of The Derivative

We routinely encounter problems in physics wherein we seek a function $g(x)$ whose derivative $g'(x)$ has been specified. We refer to $g(x)$ as the *integral* of $g'(x)$, using the notation $g(x) = \int g'(x)\,dx$, and we may work back and forth between derivatives and integrals. Second and higher order derivatives also appear in differential equations that describe a physical phenomenon such as a force. They are more difficult to solve and are often approximated.

A function may also depend on more than one variable, say x and y, with the function being written as $g(x, y)$. As an example, let us use x to represent the length of a rectangle and y its width, which is independent of x. The area of the rectangle is now x multiplied by y, or $g(x, y) = x \times y = x\,y$, and we may still take the derivative with respect to x while holding y constant. This form of derivative is called the "partial derivative of $g(x, y)$ with respect to x", and is written as $\partial g/\partial x|_y$, or more simply as $\partial g/\partial x$. We may also interchange the roles of x and y to obtain the partial of $g(x, y)$ with respect to y. Second and higher partial derivatives are similarly defined.

Vectors And Coordinates

A vector is often described simply as "a quantity which has magnitude and direction". The vectors in our analyses generally depend on the three spatial coordinates and time. However, they may also be functions of other variables such as object velocity or field strength at a location. The most common vector is the position $\underline{\mathbf{r}} = x\,\underline{\mathbf{i}}_o + y\,\underline{\mathbf{j}}_o + z\,\underline{\mathbf{k}}_o$, where $\underline{\mathbf{i}}_o$, $\underline{\mathbf{j}}_o$, and $\underline{\mathbf{k}}_o$ are unit vectors forming the three spatial axes of the selected coordinate system, and x, y, and z are the vector magnitudes in the respective directions.

The *inner or dot product* of two vectors $\underline{\mathbf{v}} = v_x\,\underline{\mathbf{i}}_o + v_y\,\underline{\mathbf{j}}_o + v_z\,\underline{\mathbf{k}}_o$ and $\underline{\mathbf{A}} = A_x\,\underline{\mathbf{i}}_o + A_y\,\underline{\mathbf{j}}_o + A_z\,\underline{\mathbf{k}}_o$ is a scalar $\underline{\mathbf{v}} \bullet \underline{\mathbf{A}} = v_x A_x + v_y A_y + v_z A_z = |\underline{\mathbf{v}}|\,|\underline{\mathbf{A}}| \cos\beta_{vA}$, where β_{vA} is the intersection angle between $\underline{\mathbf{v}}$ and $\underline{\mathbf{A}}$. The vector *cross product* $\underline{\mathbf{v}} \times \underline{\mathbf{A}}$ is another vector $|\underline{\mathbf{v}}|\,|\underline{\mathbf{A}}|\,\underline{\hat{\mathbf{u}}}_{vA} \sin\beta_{vA}$, where $\underline{\hat{\mathbf{u}}}_{vA}$ is a unit vector perpendicular to the plane formed by $\underline{\mathbf{v}}$ and $\underline{\mathbf{A}}$. The cross product is visualized by placing one's right hand on $\underline{\mathbf{v}}$ and pushing it into $\underline{\mathbf{A}}$, with the thumb pointing in the direction of $\underline{\mathbf{v}} \times \underline{\mathbf{A}}$. We also use the *curl* of $\underline{\mathbf{A}} = \nabla \times \underline{\mathbf{A}} = \underline{\mathbf{B}}$, the divergence $\nabla \bullet \underline{\mathbf{A}}$, and the gradient ∇V of an algebraic function V, called a *scalar*. Each has a physical meaning as discussed briefly in Appendix 1.

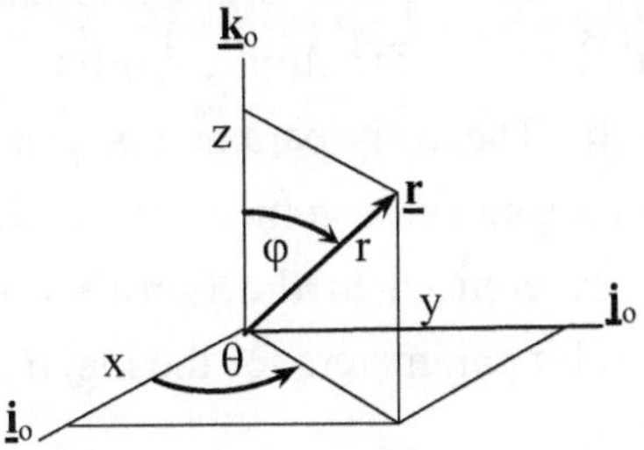

Figure 2.2. Spherical Polar Coordinate System

We define the distance r to a location as $r = |\underline{\mathbf{r}}| = (x^2 + y^2 + z^2)^{1/2}$ and use the angles θ to specify its location relative to the $\underline{\mathbf{i}}_o\underline{\mathbf{j}}_o$ plane and φ to specify the inclination of its position relative to the $\underline{\mathbf{k}}_o$ axis. These are *spherical polar coordinates*. Figure 2.2 depicts $\underline{\mathbf{r}}$ in both coordinate systems.

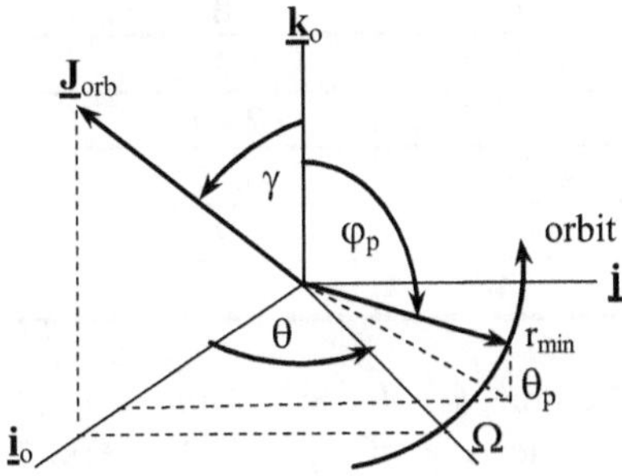

Figure 2.3. Geometry For An Inclined Orbit

The cross product of planetary position $\underline{r}$ and velocity $\underline{v}$ is the vector $\underline{J}_{orb} = \underline{r} \times \underline{v}$, depicted in Figure 2.3 as lying in the $\underline{i}_o\underline{j}_o$ plane when there is no frame rotation. It is constant in magnitude and direction, and is inclined at an angle γ to $\underline{k}_o$. $\underline{J}_{orb}$ is perpendicular to the orbital motion, but the orbit *inclination* to $\underline{k}_o$ does not necessarily lie along $\underline{J}_{orb}$. The *semi-major axis* a_n and the *eccentricity* ε (not shown) are determined by combining $\underline{J}_{orb}$ with the energy of the orbit, and they set the size and shape of the ellipse.

The value of θ at which the orbit intersects the $\underline{i}_o\underline{j}_o$ plane on its northward journey is the *ascending node* Ω. The angle θ_p where r is a minimum is the *argument of perihelion*, and is denoted by the symbol ω when measured relative to Ω. The angles ι, Ω, and ω orient the elliptical orbit in a selected $\underline{i}_o\underline{j}_o\underline{k}_o$ frame. The *true anomaly* is the planet's position in its orbit at a given time called the epoch. For a circular orbit, θ is the product of its constant angular frequency and time, but for elliptical orbits an average term called the *mean anomaly* is used. These six parameters – inclination, semi-major axis, eccentricity, ascending node, argument of perigee, and true or mean anomaly at epoch – are referenced to the Earth's orbit plane. See Appendix 2. Table 1a lists the orbit parameters for the major planets.

Table 1a. Planetary Orbit Parameters [*]
1 astronomical unit (a.u.) ≅ 1.496×10^8 kilometers

Planet	a_p (a.u.): semi-major axis	ε_p : orbit eccentricity	ι_p: orbit inclination	Ω_p: longitude of node
Mercury	0.3871	0.2056	7.00°	47.146°
Venus	0.7233	0.0068	3.40°	75.780°
Earth	1.0000	0.0168	0.00°	0.000°
Mars	1.5237	0.0933	1.85°	48.786°
Jupiter	5.204	0.0488	1.31°	100.49°
Saturn	9.582	0.0557	2.49°	113.64°
Uranus	19.189	0.0472	0.77°	74.00°
Neptune	30.071	0.0087	1.77°	131.78°
Pluto	39.264	0.2488	17.15°	110.29°

[*] The orbital data sources are Chebotarev, G., *Analytical and Numerical Methods of Celestial Mechanics*, American Elsevier Pub. Co., Inc., NY (1967), Ap. 1, 2, 3 (rounded off), epoch Greenwich mean noon 1 January 1900; and the CRC Handbook of Chemistry and Physics, CRC Press, Inc., Boca Raton, FL (1985), F-123-134.

Table 1b. Planetary Spin and Mass Parameters

Body	ν: obliquity of spin in degrees	τ: period of spin in days	μ_p: mass in Earth masses	ρ: density in gms/cc
Sun	7.0° *	24.66	332,999	1.410
Mercury	0.0°	58.82	0.053	5.431
Venus	179.9° R	234.6	0.817	5.256
Earth	23.45°	1.000	1.000	5.519
Mars	25.20°	1.029	0.107	3.907
Jupiter	3.13°	0.414	317.8	1.326
Saturn	26.73°	0.440	95.152	0.687
Uranus	97.77° R	0.718	14.536	1.27
Neptune	28.32°	0.671	17.147	1.64
Pluto	119.59° R	6.387	0.0022	2.03

* The sun's spin axis is referenced to the ecliptic. CRC Handbook. All other spin angles are referenced to the planet's orbit. The letter R indicates retrograde spin.

Table 1b shows the relative masses and densities of the planets, and the inclinations of their spin axes relative to their orbital planes. All of the planets except Mercury and Venus possess significant spins inclined to the orbit plane, *i.e.,* spin obliquity ν. † The outer planets spin rapidly, but only two terrestrial planets, Earth and Mars, have significant spin rates. The Relative Mass column shows the sun and planets masses as multiples of the Earth's mass. The terrestrial planets have densities of 3.9 to 5.5 grams per cubic centimeter, but the densities of the outer planets are less than half those of the terrestrials.

The Orbit Equation And Lagrangian Formulation

Using the derivative expressions for Newton's gravitational force $\underline{\mathbf{F}} = -\nabla V = -\mu_p M g_o \, \underline{\hat{\mathbf{u}}}_r / r^2$, where μ_p is the planet mass, M is the mass of the sun,

† The small spins of Mercury and Venus are nearly perpendicular to their orbits, but the same planet side does not always face the sun, *i.e.,* they are not *gravity-gradient stabilized.* The term refers to a small mass on a boom which stabilizes satellite attitude. The tiny difference in the gravitational force between the mass and the satellite body keeps the same side of the satellite facing the Earth. The asymmetrical mass distribution of the Moon produces gravity-gradient stabilization relative to the Earth.

and g_o is the gravitational constant, we have derived the vector components of the equations of motion for the planetary orbits in Appendix 2. In the absence of a vector potential, equation 2 would specify the planet's azimuth angle location θ as measured in the solar $\underline{\mathbf{i}}_o\underline{\mathbf{j}}_o$ plane, where r is the distance to the sun. The angular momentum vector $\underline{\mathbf{J}}_{orb}$ may be inclined to the polar axis $\underline{\mathbf{k}}_o$ at an angle γ and has a square magnitude of $J_{orb}^2 = r^2\varphi'^2 + r^2\theta'^2 \sin^2\varphi$. Orbit eccentricity and semi-major axis are denoted as ε and a_p, respectively.

$$r = a_p (1 - \varepsilon^2)/[1 + \varepsilon \cos(\theta - \theta_p)],$$
$$\text{where } a_p (1 - \varepsilon^2) = J_{orb}^2/(Mg_o) \text{ and } \cos\varphi = -\sin\iota \cos\theta, \qquad (2)$$

$$L = T - V = \mu_p (r'^2 + r^2\varphi'^2 + r^2\theta'^2 \sin^2\varphi)/2 + \mu_p Mg_o/r. \qquad (3)$$

For a closed system wherein the energy E is the sum of kinetic energy T and potential energy V, the scalar $L = T - V$ is called the *Lagrangian*, after its proponent Joseph Louis Lagrange (1736-1813). The essential feature of the theory is that the equations of motion satisfy the relation $d/dt\,(\partial L/\partial q_i') = \partial L/\partial q_i$, where q_i and $q_i' = dq_i/dt$ are generalized coordinates and their respective time derivatives, both of which may appear in an energy expression. † The Lagrangian methodology is superior to the force derivation in that it avoids vector equations and simplifies the inclusion of terms other than V.

We may use Lagrange's method with the gravitational potential energy $V(r) = -\mu_p Mg_o/r$ to derive expression 2 above. Expressing T and V in spherical coordinates and not including other possible energy terms, expression 3 provides the classical Lagrangian for the planetary orbits, where we are using the symbols $r' = dr/dt$, $\varphi' = d\varphi/dt$, and $\theta' = d\theta/dt$ for time derivatives. The partials of L with respect to r, φ, θ, and their derivatives provide the same equations of motion that one obtains from Newton's equation. We shall rely on Lagrange's approach in our gravitational field modeling, simply adding potential and kinetic energy terms as we proceed from basic Newtonian theory to more involved treatments, as in Appendix 3.

---

† The approach is an exercise in manipulating derivatives. See, *e.g.*, Becker, R. A., *Introduction to Theoretical Mechanics*, McGraw-Hill, NY (1954), pp. 322-326.

Canonical Variables And The Hamiltonian

Newton's equation is a vector equation whose components are three second-order differential equations involving six integration constants. In Cartesian coordinates the constants are three initial components of position $\underline{\mathbf{r}} = x\,\underline{\mathbf{i}}_o + y\,\underline{\mathbf{j}}_o + z\,\underline{\mathbf{k}}_o$, plus three components of velocity $\underline{\mathbf{v}} = v_x\,\underline{\mathbf{i}}_o + v_y\,\underline{\mathbf{j}}_o + v_z\,\underline{\mathbf{k}}_o$, or equivalently, three momentum components of $\underline{\mathbf{p}} = p_x\,\underline{\mathbf{i}}_o + p_y\,\underline{\mathbf{j}}_o + p_z\,\underline{\mathbf{k}}_o$, where $p_x = \mu_p v_x$, etc., and μ_p is the object's mass. In addition to the Lagrangian, we may also use a tool called the Hamiltonian, H $(q, q_2, \ldots q_k, p_1, p_2, \ldots p_k, t)$, which specifies the system energy as a function of k generalized coordinates, q_i, and their corresponding momenta $p_i = \mu_p\, dq_i/dt$. The equations of motion are derived in a manner similar to the Lagrangian, but the momentum parameters are treated as independent variables on equal footing with those of position. The result is six first-order differential equations, rather than the three second order Newtonian equations. If H is not an explicit function of time and contains only q_i variables, their values are constant. Such a set is called *canonical variables*. For solar orbits the six *canonical variables* are the semi-major axis, eccentricity, inclination, ascending node, argument of perihelion, and mean anomaly at epoch. Forces below the orbit level are treated as perturbations.

Specific Force And Specific Angular Momentum, Etc.

In the analyses that follow, we use the term "specific" as a modifier of force, angular momentum, and perhaps other quantities. In this context, the term means the force, etc., divided by the mass of the object subjected to it. This procedure is a shorthand method that simplifies many expressions.

Frame Rotations And Euler's Dynamical Equation

Newton's equation by itself in spherical coordinates does not model the effects of a non central force $\underline{\mathbf{F}}_{nc}$, *i.e.,* a force not aligned with $\underline{\mathbf{r}}$. In order to resolve this problem Leonhard Euler (1707-1783) developed a dynamical equation, wherein a torque $\underline{\boldsymbol{\tau}} = \underline{\mathbf{r}} \times \underline{\mathbf{F}}_{nc}$ modeled in an inertial frame causes the body frame to rotate with an angular frequency of $\underline{\boldsymbol{\varpi}} = (\varpi_x, \varpi_y, \varpi_z)$. Even a relatively small non-central force $\underline{\mathbf{F}}_{nc}$ can produce a significant effect on the body's motion since it is multiplied by r. Euler's equation is used to model the motions of rigid bodies such as gyroscopes and spinning tops, but it applies equally well to orbits, as indicated by equation 4, and serves to increase the angular momentum from $\mu_p\underline{\mathbf{J}}_{orb}$ to a total of $\mu_p\underline{\mathbf{J}}_{oz}$, where $\mu_p\, d\underline{\mathbf{J}}_{orb}/dt$ is usually negligible. Our analyses show that application of Euler's equation to planetary motion adequately explains many unexplained phenomena.

$$\underline{\tau} = \underline{\mathbf{r}} \times \underline{\mathbf{F}}_{nc} = \mu_p\, d\underline{\mathbf{J}}_{orb}/dt + \underline{\varpi} \times \mu_p\underline{\mathbf{J}}_{orb}. \tag{4}$$

Unfortunately many frame rotations associated with the orbital and body motions of the planets occur over very long periods of time, which increases the difficulty of validating them. For example, Neptune has completed only three orbits about the sun since the time of Copernicus, circa 1500 A.D. The period for the Earth's orbit at once per year is suited for modeling orbital frame rotations, but the periods required for the Earth and other planets to cycle through their smaller body frame rotations are much longer.

Two of these lesser frame rotation phenomena are *precession* and *nutation*, and we see them displayed in the motion of a child's top. When the top is set to spinning and released on a hard surface, its point of contact with the surface remains fixed but its inclined spin axis begins to move around in a circular motion called precession. Simultaneously, the spin axis begins to wobble, dipping up or down and then recovering, and doing so repeatedly. This is called nutation. Both phenomena are due to the force of gravity acting on the top in a direction which does not lie along the spin axis, *i.e.*, a *non central force*. The periods for precession and nutation of the spinning planets are of the order of tens to hundreds of thousands of years.

Overview Of The Moon's Orbit

Happily, the orbit of the Moon exhibits the same two phenomena over a time scale ranging from hours to about 20 years. Its orbit precession and nutation are nothing more than small frame rotations which we have modeled in spherical polar coordinates in Appendix 3 to demonstrate the validity of our approach. We note the Moon's existence has led to many other scientific discoveries, not the least of which is the law of gravitation. † Its satellite to parent planet mass ratio is about 1/81, which is much more nearly equal than that for any other major planet in the solar system. The Moon is more of a dead planet than a typical planet moon, and although the center of mass for the two lies inside the Earth's diameter, we regard the two bodies as forming a binary planetary system. In fact, it is their center of mass which revolves around the sun to provide the Earth's orbit. ††

The average of the Moon's orbital period about the Earth is 27.3 days, but the period varies as much as 7 hours due to frame rotation. Its orbit precession over a period of 18.6 years is also due to frame rotation, which causes the Moon's orbit inclination relative to the Earth's equator to vary from 18.3 to 28.6 degrees during this same period. Its inclination to the Earth's orbit vector varies by about 0.2 degree over an entirely different period of time, but is also due to a component of frame rotation. In the interest of clarity our analysis models the Moon's frame rotations in commonly used coordinates, rather than the specialized ones used by astronomers.

While they are only approximate, our results for the Moon provided in Appendix 3 are straightforward applications of frame rotation theory which a college level student of physics should be able to understand and appreciate. Not to be overlooked is the showing that an orbital torque is produced by a geometrical configuration wherein a purely central solar force is acting on a stabilized, inclined spinor. This fact is especially important when considering torques on the spinning Earth.

---

† See, Fisher, C., *The Story of the Moon*, Garden City, NY, Doubleday, Doran & Co., Inc. (1943).

†† The center of mass for the Earth-Moon system is located about two-thirds of the distance from the Earth's center to its surface. The fact that the center of the system lies inside the Earth does not preclude its logical designation as a binary planet system, probably formed by a third body impact with the Earth eons ago. The designation of "binary" based on the location of the center of mass being external to both bodies is the community's chosen nomenclature, not nature's.

3 – PROPAGATION OF GRAVITY EFFECTS BY WAVES

An understanding of the solar system requires a means of explaining the mechanism by which the sun as a field source propagates its effect on a mass μ. The explanation should also show why some field states are allowed and others are not. Newton simply stated that he could not explain the observed $1/r^2$ solar attraction. Einstein based his theory on an assumed warping of a space-time continuum. Our model envisions vacuous Euclidean space wherein the field source creates standing waves to propagate forces that interact linearly with μ. Effects of a source composed of small masses are determined to first order by their sum, *e.g.*, M for the sun. Our discussion is initially simple, but becomes more complicated for three-dimensional waves.

We envision gravitational waves as providing velocity increments to a mass μ directed toward the source, as opposed to miniscule waves which cause only slight fluctuations in the position of μ. We might compare the waves to an undertow in the surf which causes a floating object to move out to sea. Thus, an object initially at rest at a location $\underline{\mathbf{r}}$ in the solar field obtains continuous velocity increments from the waves, *i.e.*, accelerations, and eventually falls into the sun at a high speed. The end velocity is determined by the increase in kinetic energy due to the loss of potential energy, $V(r) = -\mu M g_o/r$. But if an object initially has sufficient velocity directed at a right angle to the line connecting it to the sun, its path steadily curves due to the continuous wave impulses, and the resulting motion is an orbit about the sun. Absent the waves, the object would continue along its initial path.

We gain several insights by modeling gravity waves to propagate the field. One is the result that the amplitudes of waves from a point source decrease as $1/(4\pi r^2)$ over a spherical surface at a distance r. This divergence explains the geometrical behavior of Newtonian's force $\underline{\mathbf{F}}_g = -\mu M g_o \underline{\hat{\mathbf{u}}}_r/r^2$, where $\underline{\hat{\mathbf{u}}}_r$ is a unit vector in the direction of $\underline{\mathbf{r}}$ and the gravitational constant g_o includes a factor of $1/(4\pi)$. The second is a finding that stable orbital states are based on half-integral multiples of a specific angular momentum constant σ set by the field source. We further find that the total orbit energy E is a perfect square *constant* that enables separation of the wave equation into its components. Constancy of E is usually assumed without showing necessity.

Wave Theories

Theories for classical wave motion in conduits such as strings have been well developed in the past. But unconventional wave theories have recently been proposed in an attempt to explain basic properties of matter and field interactions. Fundamental particles have given way to concepts of tiny vibrating strings with dimensions of the *Planck length*, $\ell_o = [g_o\hbar/c^3]^{1/2}$, where $\hbar = h/(2\pi)$ is the reduced Planck constant of atomic theory and c is the speed of light. Setting $g_o \cong 6.670 \times 10^{-11}$ meters cubed per kilogram per second squared with $\hbar \cong 1.0544 \times 10^{-34}$ kilogram-meters squared per second and $c \cong 2.998 \times 10^8$ meters per second, we obtain $\ell_o \cong 1.616 \times 10^{-33}$ centimeter. The tensions proposed for these conceptual strings are beyond those of any force we would ever calculate for classical observations.

Some cosmological theories are also based on the Planck length. In one proposal the entire mass of the universe is imagined to be compressed in a volume that is a tiny fraction of ℓ_o during an assumed creation of the universe via a *big bang*. The Planck length has further been related to a theory of *black holes* when the ratio of the mass of a star to the volume of space it occupies exceeds a value called the Schwarzschild limit. It has been claimed that nothing, not even a light ray, can escape a black hole. It necessarily follows that black holes are not directly observable and have been inferred with some faith and caveats. Despite contrary claims, such concepts are not validated by observations, and they often lead to hypotheses based on enormous values of negative energy required to offset comparable positive values, leaving observables which are smaller by orders of magnitude.

These fantastic flights of imagination based on the Planck length seem to have little relevance to more mundane observations for the solar system. If, however, we consider a wave theory using a different fundamental length of approximate value $a_o = 0.285 \times 10^8$ kilometers for the solar field, we obtain a much different result, with a string tension of the order of the gravitational force and wave lengths of order of the planetary spacing. We also find that the square of a planet's specific angular momentum is a quarter-fraction multiple of σ^2, where $\sigma = (Mg_oa_o)^{1/2}$ is about 1.946×10^9 kilometers squared per second, the solar field equivalent of $\hbar$ divided by the electron mass m_e.

The base length a_o that defines ɵ varies with the mass of the source M, but the form $(Mg_o/a_o)^{1/2}$ appears to remain constant for all gravitational fields at $v_g \cong 68.28$ kilometers per second. We note that the corresponding set of constants in atomic theory is $\alpha_e c$, where $\alpha_e \cong 1/137.04$ is called the *fine structure constant* and c is the speed of light. For reasons discussed below, we accept the thesis that gravity waves in a vacuum travel at light speed. If we then equate v_g to $\alpha_g c$ for a fine structure constant α_g for the gravitational field, we obtain $\alpha_g \cong 1/4393$. Why shouldn't the gravitational field have its own parameters? A long-sought unified field theory can be achieved by applying similar methodologies for the classical fields. More specifically, unification lies in modeling the effects of waves generated by the field source, rather than by forcing the same parameters onto both fields. We have adopted the position that space has no properties, or "fabric", † and is simply a region of nothingness through which gravitational and electromagnetic waves propagate at c, the observed speed of light.

One Dimensional Wave Motion

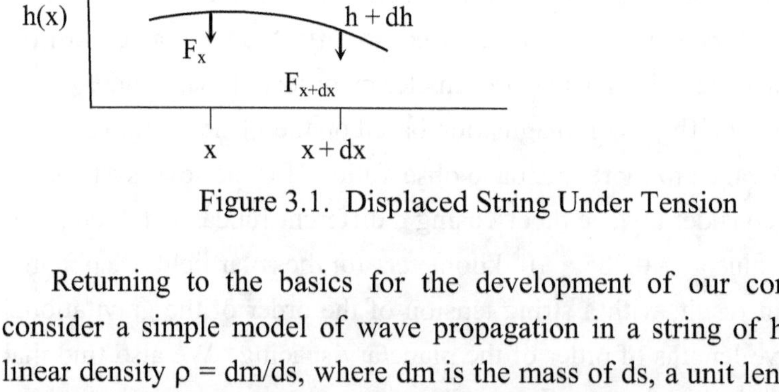

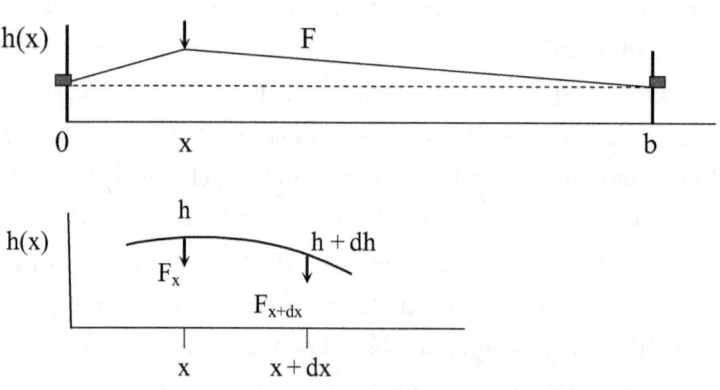

Figure 3.1. Displaced String Under Tension

Returning to the basics for the development of our concept, let us consider a simple model of wave propagation in a string of homogeneous linear density $\rho = dm/ds$, where dm is the mass of ds, a unit length of string.

† It seems that much of the physics community has come full circle from rejecting the concept of an aether to resurrecting it as a fabric for a space-time continuum.

The total length of the string is b, and it is tied to a stop at both ends under constant tension, *i.e.*, a uniform force of F along its length. When we pluck it at a point x indicated in Figure 3.1, the tension changes only slightly. Displacements of h(x) at various values of x are indicated in the figure, but as time progresses h(x) will be a function of time t as well as position x and will be written as h(x, t), which has the dimension of length.

Using the slope of h(x) to model the downward forces at the two points x and x + dx and approximating ds as dx for small disturbances, the displacement results in forces of $F_{x+dx} = F\,(\partial h/\partial x)_{x+dx}$ at x + dx and $F_x = F\,(\partial h/\partial x)_x$ at x, with a difference of $F_{x+dx} - F_x$. Midway between x and x + dx, we may express $(\partial h/\partial x)_{x+dx}$ as $(\partial h/\partial x)_{x+dx/2} + (\partial^2 h/\partial x^2)_{x+dx/2}\,(dx/2)$ and $(\partial h/\partial x)_x$ as $(\partial h/\partial x)_{x+dx/2} - (\partial^2 h/\partial x^2)_{x+dx/2}\,(dx/2)$. The force difference becomes $F_{x+dx} - F_x = F\,(\partial^2 h/\partial x^2)_{x+dx/2}\,dx$ and causes an acceleration in the mass $dm = \rho\,dx$ at the location x + dx/2, which may be expressed by Newton's law as $F_{x+dx} - F_x = dm\,(\partial^2 h/\partial t^2)_{x+dx/2}$. Dropping subscripts, we obtain equation 1.

Setting the constant F/ρ to c_s^2, which has the dimension of velocity squared, we see that *any* differentiable form $h(x - c_s t)$ or $h(x + c_s t)$ will solve equation 1. These forms are called d'Alembert's solution, and they represent waves moving through the string with velocities of $-c_s$ and $+c_s$. If we alternatively express h(x, t) as the product of separable functions of x and t, *i.e.*, $h(x, t) = \Im(t)\,\Psi(x)$, equation 1 leads to equation 2. This equation will hold for independent variations in t and x only if both sides are equal to a separation constant, which can be expressed as $-\omega_n^2/c_s^2$. We then obtain equations 3a and 3b, with the solutions given by equations 4a and 4b.

$$\partial^2 h/\partial x^2 = (\rho/F)\,\partial^2 h/\partial t^2, \quad \text{where the dimension of } h(x,t) \text{ is length.} \qquad (1)$$

$$(1/c_s^2)\,(1/\Im)\,d^2\Im/dt^2 = (1/\Psi)\,d^2\Psi/dx^2, \qquad \text{where } h(x,t) = \Im(t)\,\Psi(x), \qquad (2)$$

$$d^2\Psi/dx^2 + (\omega_n^2/c_s^2)\,\Psi = 0, \qquad (3a)$$

$$d^2\Im/dt^2 + \omega_n^2\,\Im = 0, \qquad (3b)$$

$$\Psi(x) = A_n \cos(\omega_n x/c_s) + B_n \sin(\omega_n x/c_s), \qquad (4a)$$

$$\Im(t) = C_n \cos\omega_n t + D_n \sin\omega_n t, \quad \text{where } A_n, B_n, C_n \text{ and } D_n \text{ are constants.} \qquad (4b)$$

There seems to be a conflict between d'Alembert's solution and $h(x, t) = \Im(t)\,\Psi(x)$, which is called Bernoulli's solution. However, in 1807 a French mathematician, Jean B. J. Fourier (1768-1830), showed that an analytical function $\Psi(x)$ can be decomposed over the interval $-b \leq x \leq b$ into an infinite series of the form of expression 5. Since $\int \cos(nx/b)\cos(mx/b)\,dx$ and $\int \sin(nx/b)\sin(mx/b)\,dx$ average to zero as x ranges from $-b$ to b if $m \neq n$, the terms $A_n = \int \Psi(x)\cos(nx/b)\,dx/b$ and $B_n = \int \Psi(x)\sin(nx/b)\,dx/b$ specify the amplitudes of the individual wave components. We call the resulting expression for $\Psi(x)$ a Fourier series, and it supports the contention that Bernoulli's solution is just a different form of the d'Alembert solution. †

$$\Psi(x) = A_o + A_1\cos(x/b) + B_1\sin(x/b) + \ldots + A_n\cos(nx/b) + B_n\sin(nx/b) + \ldots \qquad (5)$$

As an example of wave motion, let us suppose that we are standing in the ocean near the shore, watching the waves go by and that C_n in equation 4b is zero. Our value of x is fixed in equation 4a, and we observe the depth of the water moving up and down at our location in accordance with 4b. For a single frequency of f_n as the number of oscillations per second, the depth of the water might be roughly modeled by Figure 3.2 as a function of time.

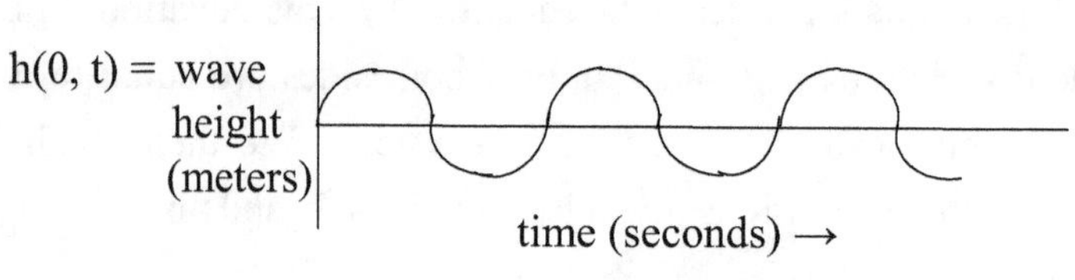

Figure 3.2. Wave Height As A Function Of Time Only

From an alternative point of view, at some given time we might look along a line running from the shore to the furthest waves, where the point x is now a variable rather than being fixed. We see that the depth, as sketched in Figure 3.3, resembles the oscillation at a fixed distance over time, but as a

† The wave equation is frequently associated with the names of Daniel Bernoulli (1700-1782), Jean d'Alembert (1717-1783), and Leonhard Euler (1707-1783).

function of position rather than time. The distance between successive peaks defines a wave length, λ_n. The product of the wave length and frequency f_n of the occurrence of peaks and valleys is the speed c_s at which the wave is traveling, *i.e.*, $c_s = \lambda_n f_n$. The frequency is related to ω_n, the mathematical *angular frequency* of the wave equation, by the relationship $\omega_n = 2\pi f_n$. Our rough example is, of course, overly simplistic since the ocean waves are two-dimensional and are eventually extinguished as they near the shore.

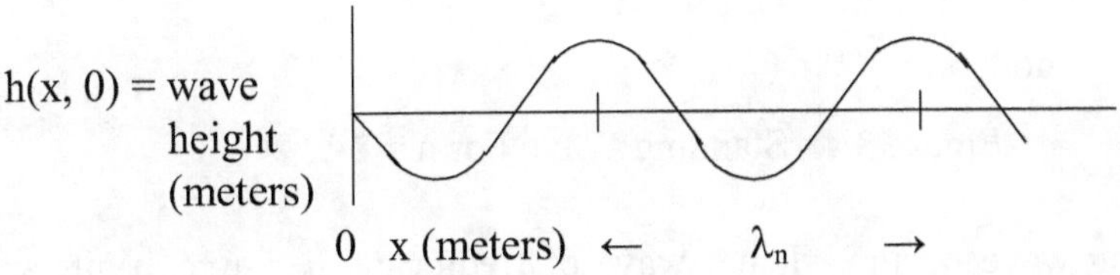

Figure 3.3. Wave Height As A Function Of Position Only

Standing Waves And The Aggregation Of Matter

If the ends of the string in our previous example are fixed, A_n in expression 4a must be zero since $h(x, t)$ is zero at $x = 0$. But $h(b, t)$ is also zero. In order for B to be non-zero in expression 4a, we must have $\omega_n b/c_s = n\pi$, where n is an integer. Expressing $C/(C^2 + D^2)^{1/2}$ as the constant $-\sin\omega_n t_o$ for some specified time t_o with $D/(C^2 + D^2)^{1/2}$ being expressed as $\cos\omega_n t_o$ in equation 4b, we may write $\Im(t)$ as $C_n \sin\omega_n(t - t_o)$, where $\omega_n t_o$ is called the phase angle and $C_n = (C^2 + D^2)^{1/2}$. Expression 6 provides the complete string solution. When n is 1 so that $\omega_1 = \pi c_s/b$, we call ω_1 the *fundamental frequency.*

$$h(x, t) = C_n \sin(n\pi x/b) \sin\omega_n(t - t_o),$$
$$\text{where } \omega_n = n\pi c_s/b, \text{ and } C_n \text{ and } t_o \text{ are constants.} \quad (6)$$

For the frequency $\omega_n = n\pi c_s/b$, the wave amplitude is zero at <u>all times</u> at $x = m\,(b/n)$, where m is an integer less than or equal to n. These locations are called the *nodes* for the nth mode. The extreme peaks and valleys occur midway between the nodes and are called *anti-nodes*. The antinodes move smoothly from an extreme to zero and then proceed to the opposite extreme as $\sin\omega_n(t - t_o)$ changes sign and increases in absolute value. The wave form is called a *standing wave*. Figure 3.4 is an illustrative sketch of a standing wave for n = 2. The general solution for h(x, t) is a sum of such terms.

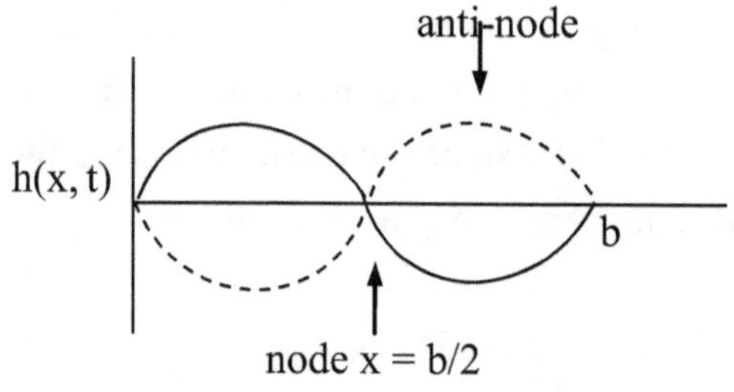

Figure 3.4. Standing Wave For n = 2

A standing wave occurs when a wave of frequency $\omega_n = n\pi c_s/b$ hits an end stop and is reflected back into the string. The reflection is superimposed on the original wave, and the process repeats when the wave arrives at the opposite stop, creating a fixed pattern. Reflected wavelengths that do not divide integrally into b interfere with the original and dissipate over time.

Let us suppose that small bits of matter are present to interact with the standing wave motion, say negligible-weight, frictionless rings free to slide along the string, or tiny floats on a water surface. The small masses will tend to aggregate at the nodes of the wave, as opposed to all other locations where they are continually buffeted about. Viewed another way, there exist regions of stability in fields propagated by standing waves where objects tend to collect and remain. However, it is not necessary that particles be present at all or any of the nodes. This simplistic one-dimensional example is representative of a process by which dispersed small particles may have accreted under their mutual gravitational attraction to form the planets.

Wave Motion In Two Dimensions

A two dimensional wave example may be based on a stretched membrane under a uniform tension of T over a surface, where the wave appears as a displacement perpendicular to the plane of the membrane. Positions in the east to west direction can be modeled as x and −x, and separately in the north to south direction as y and −y, with the edges of the membrane set by $-b_x \le x \le b_x$ and $-b_y \le y \le b_y$, as indicated in Figure 3.5.

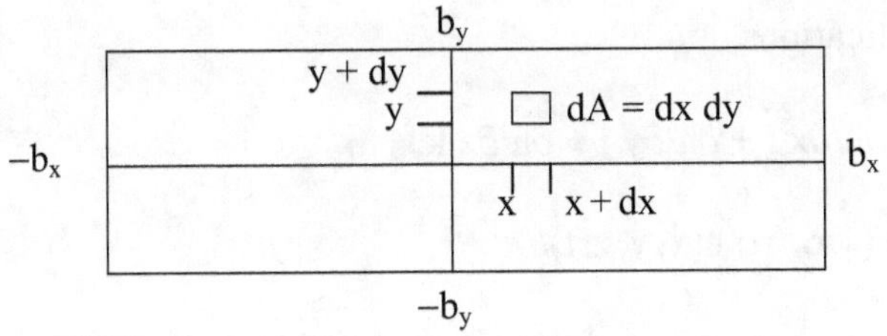

Figure 3.5. Two-Dimensional Stretched Membrane

The downward force on a small area dA is given by F = T dx dy, and changes as x or y changes over the area. However, if only y changes in response to a disturbance, the force component along the x-axis is constant at T dx, and if only x changes, the component along y remains constant at T dy. Let us write $F_{x+dx,y+dy} - F_{x,y} = (F_{x+dx,y+dy} - F_{x,y+dy}) + (F_{x,y+dy} - F_{x,y})$ and use the previous result that $(F_{x+dx,y+dy} - F_{x,y+dy}) = T\,dy\,(\partial^2 h/\partial x^2)\,dx$ and $(F_{x,y+dy} - F_{x,y}) = T\,dx\,(\partial^2 h/\partial y^2)\,dy$. The sum of the two force components then takes the form $F = (\partial^2 h/\partial x^2 + \partial^2 h/\partial y^2)\,T\,dx\,dy$. But since the mass density ρ is a constant such that $\rho\,dx\,dy = dm$, Newton's force equation $F = dm\,(\partial^2 h/\partial t^2)$ provides equation 7 for the motion of the mass dm as a function of x and y. The spatial portion of the solution for h(x, y, t) is of the form of equation 8, and the angular frequency ω_n is a vector $\underline{\boldsymbol{\omega}}_n = \omega_{nx}\,\underline{\mathbf{i}} + \omega_{ny}\,\underline{\mathbf{j}}$, with a square sum of $\underline{\boldsymbol{\omega}}_n^{\,2} = \omega_{nx}^{\,2} + \omega_{ny}^{\,2}$, reflecting independent motions along the $\underline{\mathbf{i}}$ and $\underline{\mathbf{j}}$ axes.

$$\partial^2 h/\partial x^2 + \partial^2 h/\partial y^2 = (\rho/T)\,\partial^2 h/\partial t^2, \tag{7}$$

$$h(x, y, t) = C_n \sin(n_x \pi x/b_x)\,\sin(n_y \pi y/b_y)\,\sin\omega_n(t - t_o),$$
$$\text{where } \omega_{nx} = n_x \pi c_s/b_x,\ \omega_{ny} = n_y \pi c_s/b_y, \text{ and } \omega_n^{\,2} = \omega_{nx}^{\,2} + \omega_{ny}^{\,2}. \tag{8}$$

Three Dimensional Waves

Unlike situations wherein a wave propagates perpendicularly to a surface, three-dimensional wave motion occurs within the space itself. Extrapolating from the two-dimensional example with ∇h specified by expression 9 for unit vectors of $\underline{\mathbf{i}}$, $\underline{\mathbf{j}}$, and $\underline{\mathbf{k}}$, equation 10 is the three-dimensional wave equation. We accept the wave equation as having been provided by nature, based on the fact that it correctly models observations in many physical applications. †

$$\nabla h(x, y, z, t) = \partial h/\partial x\ \underline{\mathbf{i}} + \partial h/\partial y\ \underline{\mathbf{j}} + \partial h/\partial z\ \underline{\mathbf{k}}, \qquad (9)$$

$$\nabla^2 h(x, y, z, t) = (1/c_s^2)\, \partial^2 h(x, y, z, t)/\partial t^2, \qquad (10)$$

$$h(\underline{\mathbf{r}}, \underline{\boldsymbol{\omega}}_n, t) = C_n \sin(\omega_{nx} x/c_s) \sin(\omega_{ny} y/c_s) \sin(\omega_{nz} z/c_s) \sin\omega_n(t - t_o),$$

where $-b_x \le x \le b_x$, $-b_y \le y \le b_y$, $-b_z \le z \le b_z$,
$\omega_{nx} = n_x \pi c_s/b_x$, $\omega_{ny} = n_y \pi c_s/b_y$, $\omega_{nz} = n_z \pi c_s/b_z$,
and n_x, n_y, and n_z are integers,
$\underline{\boldsymbol{\omega}}_n = \omega_{nx}\ \underline{\mathbf{i}} + \omega_{ny}\ \underline{\mathbf{j}} + \omega_{nz}\ \underline{\mathbf{k}}$,

$$h(\underline{\mathbf{r}}, \underline{\boldsymbol{\omega}}_n, t) = 0, \qquad \text{otherwise.} \qquad (11)$$

Equation 11 is the wave solution when the energy transmitted by the waves is kinetic and rigid walls restrict their space to the region $-b_x \le x \le b_x$, $-b_y \le y \le b_y$, and $-b_z \le z \le b_z$. A node occurs when a component of $\underline{\mathbf{r}} \bullet \underline{\boldsymbol{\omega}}_n/c_s$ is $m\pi$, where m is an integer less than or equal to n_x, n_y, or n_z. Using the imaginary number $i = (-1)^{1/2}$ to write $[\exp(i\, \omega_{nx} x/c_s) - \exp(-i\, \omega_{nx} x/c_s)]/(2i)$ as $\sin(\omega_{nx} x/c_s)$, where "exp" is the base of the natural logarithm, the wave solution is usually written as a sum of such exponential terms. In spherical polar coordinates $h(\underline{\mathbf{r}}, \underline{\boldsymbol{\omega}}_n, t)$ takes the form of equation 12. The variables $\underline{\mathbf{r}}$ and $\underline{\boldsymbol{\omega}}_n$ are Fourier transforms, with ∇^2 producing a negative value of ω_n^2/c_s^2 when it operates on $h(\underline{\mathbf{r}}, \underline{\boldsymbol{\omega}}_n, t)$. The units of $h(x, t)$, and thus C_n, is length.

$$h(\underline{\mathbf{r}}, \underline{\boldsymbol{\omega}}_n, t) = C_n [\exp(i\, \underline{\mathbf{r}} \bullet \underline{\boldsymbol{\omega}}_n/c_s) - \exp(-i\, \underline{\mathbf{r}} \bullet \underline{\boldsymbol{\omega}}_n/c_s)] \exp(\pm i\, \omega_n t). \qquad (12)$$

Gravity Waves In The Solar Field

A number of models exist for wave propagation in liquids and gasses based on molecular collisions and other material processes, and although we

---

† See, Lindsay, R.B., *Concepts and Methods of Theoretical Physics*, Dover Pub., Inc., New York (1951), at pp. 353-358, for additional discussion of this topic.

shall draw on these approaches, we liken gravity waves to electromagnetic wave propagation which occurs regardless of whether or not any substance is present to interact with the waves. For three dimensional applications, the form $h(\underline{\mathbf{r}}, t)$ replaces $h(x, t)$ for modeling the *wave amplitude*. If we envision gravity waves moving through a primordial cloud of dust, the wave motion should occur as longitudinal waves (rather than the transverse ones we have depicted) wherein layers of mass are alternatively compressed and rarified. However, we will not specify the precise form at this time.

The only explicit terms in the wave equation are $h(\underline{\mathbf{r}}, t)$ and c_g^2. But when we solve the equation, a vector form of $\underline{\boldsymbol{\omega}}_n$ appears in the solution due to constraints imposed for standing waves, as it does in the one dimensional example. In addition, a mass μ moving with velocity $\underline{\mathbf{v}}$ will not be in a stable field state unless its motion conforms to all of the constraints imposed by the wave equation. It must not only be located at a node, but its velocity must also be a value allowed at the node.

The three-dimensional operator $\nabla^2 = \partial^2/\partial x^2 + \partial^2/\partial y^2 + \partial^2/\partial y$ produces a negative value of ω_n^2/c_s^2 for a standing wave when it operates on $h(\underline{\mathbf{r}}, \underline{\boldsymbol{\omega}}_n, t)$ with $\underline{\mathbf{r}}$ and $\underline{\boldsymbol{\omega}}_n$ behaving as Fourier transforms. Let us suppose that we may match $\underline{\boldsymbol{\omega}}_n/c_g$ with $\underline{\mathbf{v}}_n/\sigma$ in the gravitational wave application, where $\underline{\mathbf{v}}_n$ is the allowed velocity of an object of mass μ in the nth orbital state and σ is a constant that renders $\underline{\mathbf{r}} \bullet \underline{\mathbf{v}}_n/\sigma$ dimensionless. Again using $i = (-1)^{1/2}$ to express $\sin(\underline{\mathbf{r}} \bullet \underline{\mathbf{v}}_n/\sigma)$ as $[\exp(i\, \underline{\mathbf{r}} \bullet \underline{\mathbf{v}}_n/\sigma) - \exp(-i\, \underline{\mathbf{r}} \bullet \underline{\mathbf{v}}_n/\sigma)]/(2i)$, where "exp" is the base of the natural logarithm, we obtain $-\underline{\mathbf{v}}_n^2\, \Psi/\sigma^2$ when we apply ∇^2 to $\Psi = \exp(i\, \underline{\mathbf{r}} \bullet \underline{\mathbf{v}}_n/\sigma)$. Since the kinetic energy of the mass μ is $T_n = \mu\underline{\mathbf{v}}_n^2/2$, we conclude that a mass in a stable state at the node of a simple standing wave has a kinetic energy of T_n, which appears in the form $-2T_n/(\mu\sigma^2)$ when ∇^2 operates on Ψ. We refer to Ψ as the state function for μ in the gravitational application, and use its total energy E_n, velocity $\underline{\mathbf{v}}$, and position $\underline{\mathbf{r}}$ to define the state for standing waves defined by $\underline{\boldsymbol{\omega}}_n/c_g$.

Unhappily, the form $\Psi = \exp(i\, \underline{\mathbf{r}} \bullet \underline{\mathbf{v}}_n/\sigma)$ does not solve the differential equation produced by ∇^2 when $V(r) = -\mu M g_o/r$ is included in the energy. The equation can be solved, however, by using spherical polar coordinates and expressing $\Psi(\underline{\mathbf{r}})$ as $R(r)\, \Phi(\varphi)\, \Theta(\theta)$. Using the atomic number Z of an atom, e for the electron's charge, m_e for its mass, and ε_o for the permittivity of free space, and replacing $V(r) = -\mu M g_o/r$ by $V_e(r) = -Ze^2/(4\pi\varepsilon_o r)$ and σ by $\hbar/m_e$, where $\hbar$ is the reduced Planck constant, our wave equation becomes equation 13, the Schroedinger equation of atomic physics. Its solutions are not as simple as the expression might suggest, and are overlaid with normal statistical distributions and Fermi exclusion principles due to the extremely large number of atoms and electrons involved.

$$[-\hbar^2\nabla^2/(2m_e) - Ze^2/(4\pi\varepsilon_o r)]\,\Psi = E\,\Psi. \tag{13}$$

But even with modifications, atomic physicists found that including $V_e(r)$ alone did not provide the observed energy states of some atoms, and it was necessary to include a magnetic type of potential in the energy expression. They have used the term *electron spin* to describe the phenomenon, which models an electron interaction with the magnetic field produced by its own motion. Our efforts agree that the inclusion of a magnetic term in the wave equation is necessary, but we disagree with the source modeled by atomic theorists. We have modeled a magnetic-type of flux $\underline{\mathbf{B}}_g$ which emanates from the central source, and we maintain that a similar type of flux $\underline{\mathbf{B}}$ originates in the nucleus for atomic physics applications. We treat the waves for both fields as imparting not only the force due to $-\nabla V$, but also that produced by $\underline{\mathbf{B}}_g$ and any other forces which may be present.

4 –MAGNETISM AND ITS GRAVITATIONAL EQUIVALENT

Most of us are familiar with magnets used for various purposes in everyday life and know that permanent magnets have north and south poles. If we try to force two north poles together, some strange force resists joining them, but a north pole is attracted to a south pole and vice versa. But perhaps not too many of us are aware that we create a magnetic field every time we run a current through a wire, and fewer realize that we can produce an electric current in a coil of wire just by turning the coil between two magnetic poles without touching them. The stronger the magnets and the greater the number of coils, the harder it is to turn them, but the more current we produce.

These phenomena are crucial to modern life, for we rely on coils of wire being turned in magnetic fields to create electricity to power our devices, light our homes and offices, heat and cool many of them, and cook our food. What is actually happening is that steam power or other energy sources are turning coils in strong magnetic fields to create powerful electric fields. The fields are transmitted by wires to our homes and offices at reduced levels, and we use them to push electrons through various devices and circuits.

But what does magnetism have to do with orbits? In our discussions we have indicated that nominal orbit theory applies to the microscopic world of atoms, wherein electric charges bind electrons in orbits to a nucleus. But the proper expression of orbital energy for some atoms requires the inclusion of magnetic field terms. The same type of problem occurs in modeling energy states of the solar gravitational field, but the mainstream community does not recognize the existence of such terms and incorrectly treats structured orbital states as if they were totally random. However, the structure is revealed by analogies between gravitational and electromagnetic field phenomena.

A magnetic flux $\underline{\mathbf{B}}$ is created when a collection of electrons moves in a pattern to form a current. Two wires carrying currents will experience an attendant force which causes them to be attracted to or repulsed from each other, depending on the directions of the currents. The magnitude of the force is proportional to the lengths of the wires and the strengths of the currents. If only a single wire carries a current, a charge q moving nearby with a velocity $\underline{\mathbf{v}}$ will experience a magnetic force of $\underline{\mathbf{F}}_B = q\,\underline{\mathbf{v}} \times \underline{\mathbf{B}}$, where $\underline{\mathbf{B}}$ is the magnetic flux created by the wire. The magnetic force pushes the charge in a direction perpendicular to its velocity $\underline{\mathbf{v}}$. For many applications the force $\underline{\mathbf{F}}_B$ is only a small perturbation that produces a slight variation in the charge's path. But when the charge moves in a stable orbit, the pattern of its motion creates a reinforced steady current, and its interaction with $\underline{\mathbf{B}}$ is like that of a second wire.

During the 19th century the French scientist Andre-Marie Ampere (1775-1836) conducted experiments which provide the electric current results we have cited. His findings and those of experiments by Jean-Baptiste Biot (1774-1862) and Felix Savart (1791-1841) led to expression 1 as the formulation of *Ampere's law.* Figure 4.1 illustrates the geometry.

$d\underline{\mathbf{B}} = [\kappa_o\vartheta/(4\pi)]\, d\underline{\mathbf{s}} \times \underline{\hat{\mathbf{u}}}_r/r^2$, where ϑ is the current strength, κ_o is a constant called the permeability of empty space, $d\underline{\mathbf{s}}$ is an element of the current path, $\underline{\mathbf{r}}$ is one's location in the field, $\underline{\hat{\mathbf{u}}}_r$ is a unit vector in the direction of $\underline{\mathbf{r}}$, and $d\underline{\mathbf{B}}$ is the magnetic field at $\underline{\mathbf{r}}$ created by $\vartheta\, d\underline{\mathbf{s}}$. (1)

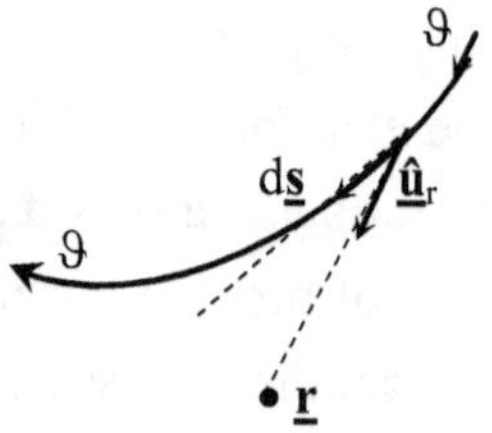

Figure 4.1. Illustration For Ampere's Law

The Source Of Gravitational Flux $\underline{\mathbf{B}}_g$

If we apply an analogy of Ampere's law to the solar gravitational field, we would expect $\underline{\mathbf{B}}_g$ to be produced by mass currents in the sun, where κ_g is the gravitational equivalent of κ_o. Expressing g_o as $1/(4\pi\varepsilon_g)$ and assuming that gravity waves travel at the speed of light c, we obtain $\kappa_g \cong 9.324 \times 10^{-27}$ meters per kilogram when we apply the analogous relationship $\varepsilon_g\kappa_g = 1/c^2$. See Chapter 10 below.

Let us model a uniform mass current ϑ flowing radially outward in the $\underline{\mathbf{i}}_o\underline{\mathbf{j}}_o$ solar plane, as in expression 2. Crossing $\vartheta\, d\underline{\mathbf{s}}$ for an element of mass flow with $\underline{\mathbf{r}}$ of expression 3 and applying Ampere's law, we obtain expression 4 for $d\underline{\mathbf{B}}_g$, which vanishes at $\varphi = \pi/2$. Expression 5 provides $\underline{\mathbf{B}}_g$, where S is the sum of the ds values. We may view ϑS as the momentum $M_{xy}v_s$ for a layer of mass M_{xy} flowing outward from the center of the sun at an average velocity v_s near the speed of light, where M_{xy} is a fraction of M and we have combined the indicated parameters into a single parameter $\sigma/2$.

$$\vartheta\, d\underline{\mathbf{s}} = \vartheta\, ds\,(\cos\theta\, \underline{\mathbf{i}}_o + \sin\theta\, \underline{\mathbf{j}}_o), \tag{2}$$

$$\underline{\mathbf{r}} = r\,(\sin\varphi\cos\theta\, \underline{\mathbf{i}}_o + \sin\varphi\sin\theta\, \underline{\mathbf{j}}_o + \cos\varphi\, \underline{\mathbf{k}}_o) = r\, \underline{\hat{\mathbf{u}}}_r, \tag{3}$$

$$d\underline{\mathbf{B}}_g = -\kappa_g\, ds\,(\vartheta\cos\varphi)\,(-\sin\theta\, \underline{\mathbf{i}}_o + \cos\theta\, \underline{\mathbf{j}}_o)/(4\pi r^2), \tag{4}$$

$$\underline{\mathbf{B}}_g = -\kappa_g\, \vartheta S\,(\cos\varphi)\, \underline{\hat{\mathbf{u}}}_\theta/(4\pi r^2) = -\sigma\,(\cos\varphi)\, \underline{\hat{\mathbf{u}}}_\theta/(2r^2), \qquad \sigma = \kappa_g\, \vartheta S/(2\pi). \tag{5}$$

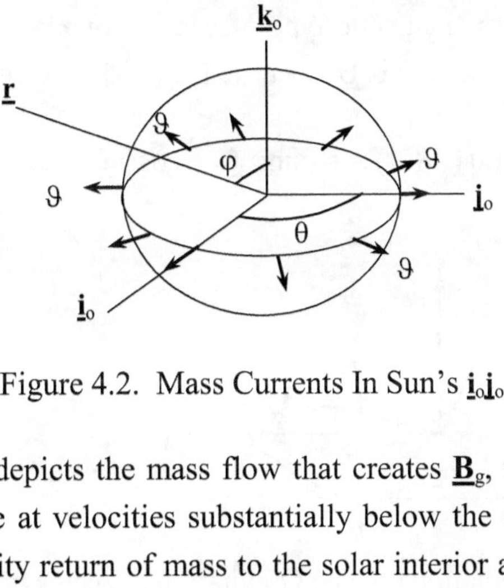

Figure 4.2. Mass Currents In Sun's $\underline{\mathbf{i}}_o\underline{\mathbf{j}}_o$ Plane

Figure 4.2 depicts the mass flow that creates $\underline{\mathbf{B}}_g$, which we interpret as being negligible at velocities substantially below the speed of light. Thus, the lower velocity return of mass to the solar interior after the outward flow is spent does not affect the value of $\underline{\mathbf{B}}_g$. An element $\vartheta\, d\underline{\mathbf{s}}_a$ adjacent to that for a given value of θ contributes a factor of $dB_{ga}\cos\alpha$ to the $\underline{\hat{\mathbf{u}}}_\theta$ component $d\underline{\mathbf{B}}_g$, where α is the-sun centered angle between $d\underline{\mathbf{B}}_g$ and $d\underline{\mathbf{B}}_{ga}$. An element of the same magnitude on the other side of θ contributes a factor of $dB_{ga}\cos\alpha$, while cancelling $dB_{ga}\sin\alpha$, and so forth. For example, at $\theta = \pi/2$ only current elements of $\vartheta\, ds$ with a component along $\underline{\mathbf{i}}_o$ contribute to $\underline{\mathbf{B}}_g$, and the contributions increase from zero as the elevation angle $\pi/2 - \varphi$ increases.

To first order, an object exterior to the sun moving with velocity v_s in the $\underline{\mathbf{i}}_o\underline{\mathbf{j}}_o$ plane, where $\cos\varphi$ is zero, will not be affected by $\underline{\mathbf{B}}_g$, but will be somewhat affected by a small dipole flux $\underline{\mathbf{B}}_{g1} = [-\kappa_g M_{xy} v_s \hat{\underline{\mathbf{u}}}_\theta/(4\pi)]\,[-\underline{\mathbf{p}}_1 \bullet \nabla(1/r^2)] = -\,\sigma\,(p_1 \sin\varphi)\,\hat{\underline{\mathbf{u}}}_\theta/r^3$ due to the separation of current elements. $M_{xy}\,\underline{\mathbf{p}}_1$ is the dipole moment, where M_{xy} is mass flowing in the $\underline{\mathbf{i}}_o\underline{\mathbf{j}}_o$ plane and the vectored separation $\underline{\mathbf{p}}_1$ is expressed in units of length. We will ignore $\underline{\mathbf{B}}_{g1}$ except in instances when an orbit lies in the $\underline{\mathbf{i}}_o\underline{\mathbf{j}}_o$ plane, as in Chapter 8 where we consider the perihelion advance of Mercury's orbit.

The Gravitational Vector Potential $\underline{\mathbf{A}}_g$

Similar to the ability to express Newton's force $\underline{\mathbf{F}}$ as $-\nabla V$, we express $\underline{\mathbf{B}}$ in electromagnetic theory as the curl of a *vector potential* $\underline{\mathbf{A}}$, *i.e.*, $\underline{\mathbf{B}} = \nabla \times \underline{\mathbf{A}}$. Like V, the vector potential $\underline{\mathbf{A}}$ has a 1/r behavior for a point source, but has the dimension of velocity and depends on angular variables as well as r. We will model an analogous magnetic type of vector potential $\underline{\mathbf{A}}_g$ for the solar field, expression 6, whose curl is $\underline{\mathbf{B}}_g$ specified above by expression 5.

$$\underline{\mathbf{A}}_g = \sigma\,(\cos\theta\,\underline{\mathbf{i}}_o + \sin\theta\,\underline{\mathbf{j}}_o)/(2r) = \sigma\,(\sin\varphi\,\hat{\underline{\mathbf{u}}}_r + \cos\varphi\,\hat{\underline{\mathbf{u}}}_\varphi)/(2r). \tag{6}$$

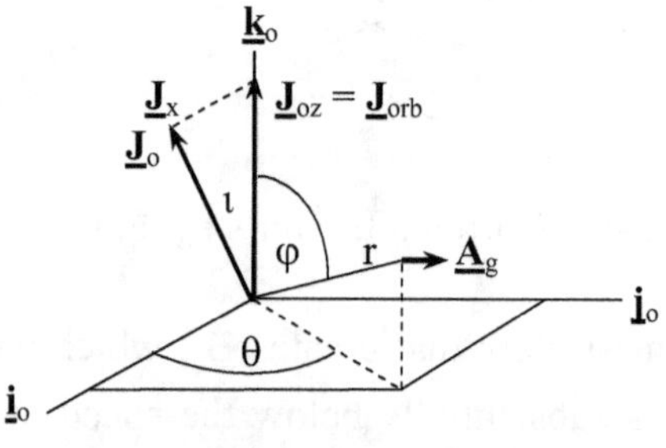

Figure 4.3. Orientation Of The Gravitational Vector Potential $\underline{\mathbf{A}}_g$

The potential $\underline{\mathbf{A}}_g$ occurs with a magnitude of $\sigma/(2r)$ in the $\underline{\mathbf{i}}_o\underline{\mathbf{j}}_o$ plane, which is normal to the polar axis $\underline{\mathbf{k}}_o$ of the $\underline{\mathbf{i}}_o\underline{\mathbf{j}}_o\underline{\mathbf{k}}_o$ solar field, as depicted in Figure 4.3. Based on our fitting of orbital data, we have found σ to be about 1.946×10^9 kilometers squared per second, *i.e.*, units of *specific angular momentum*, and it may be expressed as $\sigma = (M g_o a_o)^{1/2}$, where M is the total solar mass, g_o is the universal gravitational constant, and a_o is a length parameter for the field. When $\underline{\mathbf{A}}_g$ is added to the velocity $\underline{\mathbf{v}}$ in the expression of energy, it causes an orbit to be tilted away from the $\underline{\mathbf{k}}_o$ axis by an angle ι, as indicated in Figure 4.3 where $\mu_p\underline{\mathbf{J}}_{orb}$ is the total angular momentum lying along $\underline{\mathbf{k}}_o$. We have allowed for $\underline{\mathbf{A}}_g$ and $\underline{\mathbf{B}}_g$ not because of some abstraction, but rather, because they explain the observed orbital parameters for the planets.

Modeling The Effects Of $\underline{\mathbf{A}}_g$ On The Planetary Orbits

We include $\underline{\mathbf{A}}_g$ in the orbital kinetic energy T in the same manner as in electromagnetism, *i.e.*, $T = \mu_p(\underline{\mathbf{v}} + \underline{\mathbf{A}}_g)^2/2 = \mu_p(\underline{\mathbf{v}}^2 + 2\underline{\mathbf{v}} \bullet \underline{\mathbf{A}}_g + \underline{\mathbf{A}}_g^2)/2$, for a planet of mass μ_p. For circular orbits, $\underline{\mathbf{v}} \bullet \underline{\mathbf{A}}_g$ is zero for orbits lying in the $\underline{\mathbf{i}}_o\underline{\mathbf{j}}_o$ plane, and averages to zero for an inclined orbit. For modest eccentricities the coupling of $\underline{\mathbf{A}}_g$ with $\underline{\mathbf{v}}$ is a perturbation. Thus, even though $\underline{\mathbf{A}}_g$ varies as a function of position, it adds a constant factor of $\underline{\mathbf{J}}_x^2 = \sigma^2/4$ to $\underline{\mathbf{J}}_{oz}^2$ in the expression of orbital energy. We interpret the effect of $\underline{\mathbf{A}}_g$ as creating a composite vector $\underline{\mathbf{J}}_o$ of square magnitude $J_o^2 = J_{oz}^2 + \sigma^2/4$ inclined to $\underline{\mathbf{k}}_o$ at an angle ι, indicated in Figure 4.3. In the absence of frame rotation the orbital angular momentum vector $\mu_p\underline{\mathbf{J}}_{orb}$ lies along the polar axis $\underline{\mathbf{k}}_o$ of the field, *i.e.*, $\underline{\mathbf{J}}_{oz} = \underline{\mathbf{J}}_{orb}$, as is also observed for the principal moons of Jupiter and Saturn, whose fields do not exhibit $\underline{\mathbf{A}}_g$ components.

In Appendix 4 we address the Lagrangian L_A which includes $\underline{\mathbf{A}}_g$ without orbital frame rotation, expression 7, and derive the orbital equations of motion in polar coordinates. The only effective differences produced by the inclusion of $\underline{\mathbf{A}}_g$ in such orbits are (1) causing the observed orbit vector $\underline{\mathbf{J}}_o$ to be tilted to $\underline{\mathbf{k}}_o$ at the angle ι given by expression 8, and (2) the appearances of two perturbation terms, one of which is negligible for most orbits. The orbital angular momentum and the solution for r are essentially unchanged. For body-based geometries wherein the orbital angular momentum vector $\mu_p\underline{\mathbf{J}}_{orb}$ is not aligned with $\underline{\mathbf{k}}_o$, a current interaction with $\underline{\mathbf{B}}_g$ occurs, and the angular momentum configuration is more involved. This is the situation for the Earth's orbit, addressed in the next chapter.

$$L_A = \mu_p(\underline{\mathbf{v}} + \underline{\mathbf{A}}_g)^2/2 - V, \tag{7}$$

$$\tan\iota = \sigma/(2J_o). \tag{8}$$

5 – TORQUE ON THE ORBITS OF THE EARTH AND MARS

Just as a magnetic flux $\underline{\mathbf{B}}$ produces a torque on an inclined current loop, $\underline{\mathbf{B}}_g$ produces a torque on an orbit inclined to its polar axis. Absent stabilized spin, the orbit vector $\mu_p\underline{\mathbf{J}}_{orb}$ will be aligned with $\underline{\mathbf{k}}_o$, *i.e.,* $\cos\varphi = 0$, and $\underline{\mathbf{B}}_g$ does not impact the motion except for perturbations. This is the situation for Mercury and Venus. But when $\underline{\mathbf{J}}_{orb}$ is inclined to $\underline{\mathbf{k}}_o$ in a spin-based frame, such as exists for the Earth and Mars, a torque occurs due to the interaction of $\underline{\mathbf{B}}_g$ with the moving planet mass μ_p. The torque is offset by a component of angular momentum, similar to the offset of gravitational force by orbital angular momentum. We model the phenomenon as principal frame rotation of frequency ξ' about a $\underline{\mathbf{k}}$ axis, which produces a vector $\mu_p\underline{\mathbf{J}}_{sum}$ of magnitude $I_z(W^2 + 2W\xi' \cos\gamma + \xi'^2)^{1/2}$, where W is the average angular frequency of the orbit, a_n is its average radius, and $I_z \cong \mu_p a_n^2$. See Appendix 5 for details.

The body-based Lagrangian L_{sum} is given by expression 1, *ibid.*, whose first line terms are unchanged from L_A above. However, a potential term $\mu_p Q$ now appears to produce $a_n^2\xi'(\cos\gamma)/r$, which couples with orbital velocity of $r\,\theta'$ prior to taking the square, as $a_n^4\xi'^2(\sin^2\gamma)/(2r^2)$ stands alone. The expressions for $J_o^2 = J_{oz}^2 + \sigma^2/4$ and the angle ι are those for L_A, but the orbit averages of $\mu_p J_{orb} = \partial L_{sum}/\partial\xi'$ and $\mu_p J_{oz} = \partial L_{sum}/\partial\theta'$, equations 2 and 3, are affected. Equation 4, provided by $\partial L_{sum}/\partial\gamma$, models the torque that produces ξ'.

$$L_{sum}/\mu_p = [r' + (\sigma/2)(\sin\varphi)/r]^2/2 + [r\,\varphi' + (\sigma/2)(\cos\varphi)/r]^2/2 + Mg_o/r$$
$$+ [r\,\theta' + a_n^2(\xi'\cos\gamma)/r]^2(\sin^2\varphi)/2 + a_n^4\xi'^2(\sin^2\gamma)/(2r^2) - Q, \quad (1)$$

$$I_z(W\cos\gamma + \xi') = \text{constant} = I_z W_o = \mu_p J_{orb}, \quad (2)$$

$$I_z(W + \xi'\cos\gamma) = \text{constant} = \mu_p J_{oz}, \quad (3)$$

$$I_z W\,\xi'\sin\gamma = -\mu_p\,\partial Q/\partial\gamma. \quad (4)$$

Figure 5.1 shows the angular momentum components as they appear in the $\underline{\mathbf{i}}_o\underline{\mathbf{j}}_o\underline{\mathbf{k}}_o$ and $\underline{\mathbf{ijk}}$ frames. The vector $\mu_p\underline{\mathbf{J}}_{sum}$ is inclined to $\underline{\mathbf{k}}_o$ in the $\underline{\mathbf{j}}_o\underline{\mathbf{k}}_o$ plane of the inertial frame at an angle δ = arc-tan$[(\xi'\sin\gamma)/(W + \xi'\cos\gamma)]$. The component $\mu_p\underline{\mathbf{J}}_{oz}$ is the projection of $\mu_p\underline{\mathbf{J}}_{sum}$ on $\underline{\mathbf{k}}_o$. The body vector $\mu_p J_{orb}\,\underline{\mathbf{k}}$

aligns with $\underline{\mathbf{J}}_{orb}$ in inertial space only at a specific value of γ. The $\underline{\mathbf{i}}$ axis initially aligns with $\underline{\mathbf{i}}_o$ but rotates with ξ'. The $\underline{\mathbf{j}}$ axis is normal to $\underline{\mathbf{i}}$ and $\underline{\mathbf{k}}$.

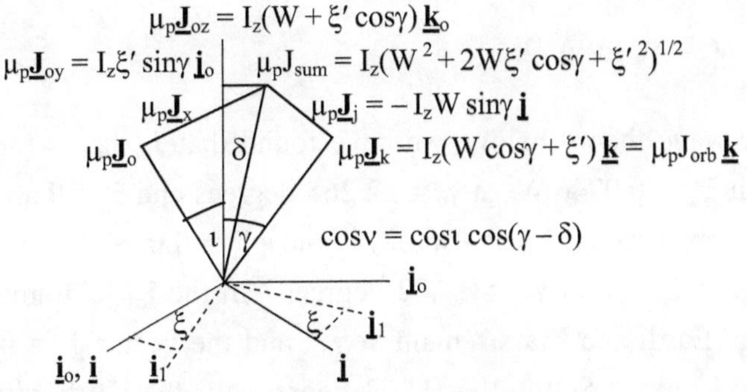

Figure 5.1. Angular Momentum Components For The Earth And Mars

Choosing the $\underline{\mathbf{ijk}}$ body frame so that $\underline{\mathbf{J}}_k$ remains aligned with the spin axis $\underline{\mathbf{\check{s}}}$ allows us to model planet body and orbital motions simultaneously in Chapter 7 below. The $\underline{\mathbf{ijk}}$ frame is called an *accelerated body frame*. As we have shown in Appendix 3 for the Moon's orbit, small torques on spinors inclined at an angle γ to the orbit create precession in the body frame and cause the inclination angle γ to fluctuate about a median value. The orbital angular momentum terms $\underline{\mathbf{J}}_j$ and $\underline{\mathbf{J}}_{sum}$ vary in the accelerated $\underline{\mathbf{ijk}}$ frame, as does the direction of $\underline{\mathbf{J}}_k$, and it is only when $\underline{\mathbf{J}}_k$ and $\underline{\mathbf{\check{s}}}$ align with $\underline{\mathbf{J}}_{orb}$ in inertial space that $\underline{\mathbf{J}}_j$, $\underline{\mathbf{J}}_k$, and $\underline{\mathbf{J}}_{sum}$ represent orbit-level vectors. $\underline{\mathbf{J}}_{oz}$ remains constant in magnitude and direction in both frames, and the magnitude of $\underline{\mathbf{J}}_k$ remains constant at J_{orb} as γ varies, although $\underline{\mathbf{J}}_k$ and $\underline{\mathbf{J}}_{orb}$ are not aligned. $\underline{\mathbf{J}}_{orb}$ remains fixed in magnitude and direction in the $\underline{\mathbf{i}}_o\underline{\mathbf{j}}_o\underline{\mathbf{k}}_o$ frame, and the body motions appear as small components of angular momentum therein.

Variations in γ in a non-inertial body frame require W and ξ' to vary in order to maintain the constancies of equations 2 and 3. Knowing the value of J_{oz}/J_{orb} allows us to solve for W and ξ' as functions of γ and W_o, equations 5 and 6. See Appendix 5. The values of $W = W_o$ and $\xi' = \xi'_o$ occur at a special angle γ_o, and there exists another angle γ_{oz} where ξ' becomes zero as ξ'_o is aliased into W_{oz}. *Ibid.* $\underline{\mathbf{J}}_{sum}$ then aligns with $\underline{\mathbf{J}}_{oz}$, as $\mu_p\underline{\mathbf{J}}_j$ becomes $\mu_p\underline{\mathbf{J}}_{fr} = -I_z W_{oz} \sin\gamma_{oz}\, \underline{\mathbf{j}}$ and $\mu_p\underline{\mathbf{J}}_k$ becomes $\mu_p\underline{\mathbf{J}}_{orb}$, with the angle δ becoming zero. The inertial configurations for the orbits of the Earth and Mars occur at γ_{oz}.

$$W = W_o (J_{oz}/J_{orb} - \cos\gamma)/\sin^2\gamma, \quad (5)$$

$$\xi' = W_o - W \cos\gamma, \quad (6)$$

$$\mu_p Q = -\mu_p r^2 (W_o - W \cos\gamma)^2/2. \quad (7)$$

Based on our wave theory below, we have found that J_{oz}/J_{orb} is $(8/7)^{1/2}$ for the Earth, with $\xi'_o \cong 0.0746\ W_o$ at $\gamma_o \cong 22.265$ degrees and $\xi' = 0$ at $\gamma_{oz} \cong 20.706$ degrees. For Mars J_{oz}/J_{orb} is $(35/32)^{1/2}$ and $\xi'_o \cong 0.0481\ W_o$ at $\gamma_o \cong 17.851$ degrees, with $\xi' = 0$ at $\gamma_{oz} \cong 17.02$ degrees. In the $\underline{\mathbf{i}}_o\underline{\mathbf{j}}_o\underline{\mathbf{k}}_o$ frame the values of γ for the Earth and Mars remain at γ_{oz}, and the vector $\underline{\mathbf{J}}_x = \sigma\, \underline{\mathbf{i}}_o/2$ connects $\underline{\mathbf{J}}_o$ to the $\underline{\mathbf{k}}_o$ axis. Setting $\iota = 11.31$ degrees for the Earth with an observed spin obliquity of $\nu_e = 23.45$ degrees and applying $\cos\iota \cos\gamma_e = \cos\nu_e$ at the present time, we obtain $\gamma_e \cong 20.68$ degrees $\cong \gamma_{oz}$. For Mars, $\iota = 9.46$ degrees with $\nu_m \cong 25.2$ degrees at the present time, so that $\gamma_m \cong 23.47$ degrees, which differs significantly from its inertial angle $\gamma_{oz} \cong 17.0$ degrees.

In Appendix 5 we have used the Lagrangian expression for $\mu_p\, \partial Q/\partial\gamma$ to provide an analogy for its components relative to classical phenomena. The potential $\mu_p Q$ is specified by expression 7, which becomes $-I_z\xi'^2/2$ when r is averaged. The Q and ξ' terms in the equation for r sum to zero when averaged over the orbit, and both are zero at γ_{oz}, as W becomes $W_o/\cos\gamma_{oz}$.

In summary, the computation of angular momentum for spin-stabilized planets involves modeling a body-based $\underline{\mathbf{ijk}}$ frame inclined at an angle γ to the solar field's $\underline{\mathbf{k}}_o$ axis. The attendant torque is produced by $\underline{\mathbf{B}}_g$ and requires a sum vector of $\mu_p\underline{\mathbf{J}}_{sum} = I_z[\xi' \sin\gamma\, \underline{\mathbf{i}} + (W + \xi' \cos\gamma)\, \underline{\mathbf{k}}]$ to offset its effects. The orbit configuration is stable in inertial space at γ_{oz}, where the square magnitude of $\underline{\mathbf{J}}_{sum}$ in the body frame becomes $\underline{\mathbf{J}}_{oz}^2 = \underline{\mathbf{J}}_{orb}^2 + \underline{\mathbf{J}}_{fr}^2$. The vector potential $\underline{\mathbf{A}}_g = \sigma(\sin\varphi\, \underline{\hat{\mathbf{u}}}_r + \cos\varphi\, \underline{\hat{\mathbf{u}}}_\varphi)/(2r)$ couples with the $\underline{\hat{\mathbf{u}}}_r$ and $\underline{\hat{\mathbf{u}}}_\varphi$ components of velocity, and adds a factor of $\mu_p\sigma^2/(4r^2)$ to the orbital kinetic energy. See Chapter 4. The end orbit vector has a square magnitude of $\underline{\mathbf{J}}_o^2 = \underline{\mathbf{J}}_{oz}^2 + \sigma^2/4$, and is inclined at an angle ι in the $\underline{\mathbf{i}}_o\underline{\mathbf{k}}_o$ plane. Neither $\mu_p\sigma/2$ nor $I_z\xi'$ is a component of orbital angular momentum, although the observed orbit vector lies in the direction of $\underline{\mathbf{J}}_o$. In Chapter 7 we will show that the $\underline{\mathbf{i}}$-axis of the $\underline{\mathbf{ijk}}$ body frame steadily precesses and the spin axis $\underline{\check{\mathbf{s}}} = \check{s}\, \underline{\mathbf{k}}$ nutates about $\underline{\mathbf{i}}$, while the angular momentum configuration remains stable in $\underline{\mathbf{i}}_o\underline{\mathbf{j}}_o\underline{\mathbf{k}}_o$ inertial space.

6 – THE NON RANDOM PLANETARY ORBIT PARAMETERS

Our gravitational results above model the effects of $\underline{\mathbf{A}}_g$ and $\underline{\mathbf{B}}_g$ but do not explain the distinct patterns of observed orbit parameters. For that we must turn to the application of the wave equation to the solar gravitational field, as specified in Appendix 6. The theory is mathematics intensive, and we now simply show that the mean orbit radii and inclinations of the planetary orbits to the solar axis reflect the presence of $\underline{\mathbf{A}}_g$ and $\underline{\mathbf{B}}_g$, and of standing waves.

Summarizing the theory, we find that the squares of specific orbital angular momenta for the major planets take the form $\underline{\mathbf{J}}_{orb}^2 = (m - 2s)(m + 1)\sigma^2$, where $\sigma \cong 1.946 \times 10^9$ kilometers squared per second, m is an integer or half-integer, and s is a rational number such that $\underline{\mathbf{J}}_{orb}^2$ is a quarter integer multiplying σ^2. Although $\underline{\mathbf{J}}_{orb}$ is aligned with the polar axis $\underline{\mathbf{k}}_o$ of the solar field in inertial space, frame rotation of $\underline{\mathbf{J}}_{fr}$ occurs when we model the orbit in a spin-based planetary frame inclined to $\underline{\mathbf{k}}_o$ in the $\underline{\mathbf{i}}_o\underline{\mathbf{k}}_o$ plane. The orbit interacts with $\underline{\mathbf{B}}_g$ to produce $\underline{\mathbf{J}}_{fr}$ and the non-zero values of s. But in inertial space the sum of $\underline{\mathbf{J}}_{orb}$ and $\underline{\mathbf{J}}_{fr}$ remains constant at $\underline{\mathbf{J}}_{oz}$ lying along $\underline{\mathbf{k}}_o$.

In all cases $\underline{\mathbf{A}}_g$ produces an angular momentum-like vector $\underline{\mathbf{J}}_x$ which lies along a field axis $\underline{\mathbf{i}}_o$ whose direction is perpendicular to the plane formed by $\underline{\mathbf{J}}_{orb}$ and $\underline{\mathbf{J}}_{oz}$. We have modeled $\underline{\mathbf{J}}_x$ by the parameter s_A, which must be a multiple of 1/2. It is 1/2 for the terrestrial planets, but increases for the outer planets as the specific angular momentum occurs in increasing multiples of σ^2. The inclusion of $\underline{\mathbf{J}}_x$ produces the observed orbit vector $\underline{\mathbf{J}}_o$ whose magnitude satisfies $J_o^2 - J_x^2 = J_{oz}^2$, and causes $\underline{\mathbf{J}}_o$ to be inclined to $\underline{\mathbf{k}}_o$ at the angle ι = arc-tangent(J_x/J_o). The planetary orbits are defined to first order by the allowed values of m, s, and s_A.

The Planetary Orbit Radii And The Titius-Bode Law

We will compare our theory results with observations based on accepted values of the sun's mass M, the gravitational constant g_o, a given planet's specific orbital angular momentum J_{orb}, and its semi-major axis a_n. We have approximated the mean orbit radius as a_n for the nth planet and ignored effects of orbit eccentricity ε, which enters the mean orbit radius computation only as ε^2. Equations 1 and 2 of our theory apply to the terrestrial planets, where we have expressed σ^2 as $Mg_o a_o$ for a base field length of a_o. The lack of significant spin for Mercury and Venus indicates that orbit-level frame rotation is not present, so their s-values should be zero. Values for the Earth and Mars are $s = 1/2$, while $s_A = 1/2$ for all of the terrestrial planets. †

$$Mg_o a_n = J_{orb}^{\ 2} = (m - 2s)(m + 1)\sigma^2 = [j(j+1) - s(s+1)]\, Mg_o a_o, \qquad \text{where } j \equiv m - s, \qquad (1)$$

$$a_n/a_o = 8/4,\ 15/4,\ 21/4,\ 32/4, \qquad \text{for } j = 2/2,\ 3/2,\ 4/2,\ 5/2, \text{ and } s = 0,\ 0,\ 1/2,\ 1/2, \text{ in order for Mercury, Venus, Earth, and Mars.} \qquad (2)$$

Normalizing to the average Earth orbit radius of one astronomical unit (a.u.) $\cong 1.496 \times 10^8$ kilometers, the observed ratios of a_n/a_o in order are 0.387, 0.723, 1.000, and 1.524 to the third decimal level. The theory values are 0.381, 0.714, 1.000, and 1.524, in good agreement in view of our approximations. The arbitrary ratios provided by the Titius-Bode law are 0.4, 0.7, 1.0, and 1.6. From Mars to Jupiter there is a gap, with available states not occupied. We conclude that regular disruptions of the region by Jupiter's strong gravitational field have prevented the formation of major planets in an otherwise stable region and have kept the mass dispersed as asteroids.

The outer planetary states involve different types of relationships and values of s and s_A other than 1/2. Jupiter is in a unique state defined by a negative value of $s = -1/4$ with $m = 9/2$ and $s_A = 1$. States based on multiples of $\sigma/2$ are also allowed, with $J_{orb}^{\ 2}$ becoming $(2m - 4s)(2m + 1)\sigma^2/4 = (m - 2s)(m + 1/2)\sigma^2$. In addition, frame rotations of $J_{fr}^{\ 2} = 4s(2m + 1)\sigma^2/4$, rather than $s(s + 1)\sigma^2$, may occur. Saturn is in a state of this form, with $m = 15/2$, $s_A = 3/2$, and $s = 5/8$. Uranus is in the simple state $m = 19/2$, $s_A = 2$, and $s = 0$. Its spin lies nearly in its orbital plane and does not reflect the

---

† We note that orbit eccentricity specified by $\varepsilon = [1 - J_{orb}^{\ 2}/(Mg_o a_n)]^{1/2}$ is observed to change slightly over periods of many thousands of years due to the gravitational attractions of neighboring planets. See, Berger, A., and Loutre, M., "Insolation values for the climate of the last 10 million years," *Quaternary Science Reviews*, **10**: 297-317, 1991. For example, ε is estimated to vary from 0.0034 to 0.058 for the Earth.

presence of frame rotation. Neptune is in the state $m = 14$, $s_A = 3$, $s = 13/8$ with $J_{orb}^2 = (m-2s)(m+1)\sigma^2$ and $J_{fr}^2 = 2s(m+1)\sigma^2$. Table 2 compares the theory results with observations for the mean orbit radii of the major planets. We attribute the gaps in m to the strong fields of the outer planets in regions adjacent to their orbits, and also surmise that the rings of Saturn reflect gravity tugs of war by Jupiter on otherwise stable satellite states in Saturn's gravitational field.

Table 2. Data Vs. Theory: Mean Orbit Radii In Astronomical Units (a.u.)

Planet	Theory: m-value	Theory: s_A-value	Theory: s-value	Theory: orbit radius (a.u.)	Data: orbit radius (a.u.)
Mercury	2/2	1/2	0	0.381	0.387
Venus	3/2	1/2	0	0.714	0.723
Earth	5/2	1/2	1/2	1.00	1.00
Mars	6/2	1/2	1/2	1.52	1.52
Jupiter	9/2	1	−1/4	5.24	5.20
Saturn	15/2	3/2	5/8	9.52	9.58
Uranus	19/2	2	0	19.00	19.19
Neptune	28/2	3	13/8	30.71	30.07

Orbit Inclinations For The Planets

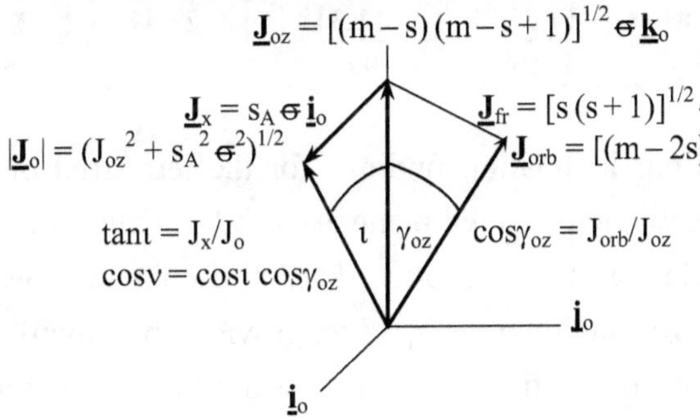

Figure 6.1. Angular Momentum Diagram For The Earth And Mars

Validation of the theory requires fitting not only the planets' semi-major axes but also their observed orbit inclinations to $\underline{k}_o$, the polar axis of the solar field. Our theory indicates that the Earth's orbit is inclined to $\underline{k}_o$ at an angle $\iota \cong 11.31$ degrees. Its angular momentum configuration is depicted in Figure 6.1 for m = 5/2 and $s_A = s = 1/2$, where the Earth's spin axis $\underline{\check{s}}$ is inclined at $\gamma \cong \gamma_{oz}$. The same diagram form applies to Mars. It also applies to Mercury and Venus with $\underline{\mathbf{J}}_{fr}$ set to zero and $\underline{\mathbf{J}}_{orb}$, when unperturbed, lying along the $\underline{\mathbf{k}}_o$ axis.

Equation 3 provides the angle ι_{sp} between $\underline{\mathbf{k}}_o$ and any planet's orbit vector $\underline{\mathbf{J}}_{op}$ measured from a suitable reference frame, including the $\underline{\mathbf{i}}_e\underline{\mathbf{j}}_e\underline{\mathbf{k}}_e$ ecliptic frame. In Appendix 5 we have estimated $\Omega_{s\pm}$, the nodes for the solar $\underline{\mathbf{i}}_o\underline{\mathbf{j}}_o$ plane, by fitting orbit inclination data for the terrestrials. Our results provide Ω_s as a line running from $\Omega_{s-} \cong 42$ degrees to $\Omega_{s+} \cong 222$ degrees in the $\underline{\mathbf{i}}_e\underline{\mathbf{j}}_e\underline{\mathbf{k}}_e$ ecliptic frame. We have used Ω_{s-} for orbits interior to the Earth's and Ω_{s+} for the exterior ones, based on the hyperbolic relationship between $\underline{\mathbf{J}}_o$ and $\underline{\mathbf{k}}_o$.

$$\underline{\mathbf{k}}_o \bullet \underline{\mathbf{J}}_{op}/J_{op} = \cos\iota_{sp} = \sin\iota_s \sin\iota_p \cos(\Omega_p - \Omega_{s\pm}) + \cos\iota_s \cos\iota_p. \qquad (3)$$

Table 3. Data Vs. Theory: Angle Results For The Terrestrial Planets
Solar Frame Parameters: $\iota_s = 11.31°$, $\Omega_{s-} = 42°$, $\Omega_{s+} = 222°$

Planet	Data: Ω_p	Data: ι_p	Data: ι_{sp} to polar $\underline{\mathbf{k}}_o$	Theory: ι_{sp} to polar $\underline{\mathbf{k}}_o$	Theory: ν (spin avg)	Obser: ν (present)
Mercury	47.146°	7.00°	18.29°	18.43°	0°	0°
Venus	75.780°	3.40°	14.27°	14.04°	0°	0.1°
Earth	n/a	n/a	11.31°	11.31°	23.47°	23.45°
Mars	48.786°	1.85°	9.46°	9.46°	19.40°	25.20°

Table 3 summarizes our inclination findings for the terrestrial planets. The two data columns show the observed planetary orbit longitudes of node Ω_p and inclinations ι_p relative to the ecliptic. The third column shows the computed values of the orbit inclinations ι_{sp} to $\underline{\mathbf{k}}_o$ provided by inputting the observed inclination values to equation 3 for our best fit of the two estimated values of Ω_s. The fourth column shows the inclinations to $\underline{\mathbf{k}}_o$ predicted by the theory, and the fifth column provides the base spin obliquity angle ν when $\underline{\check{s}}$ and $\underline{\mathbf{J}}_{orb}$ are aligned. The last column shows the present values of ν.

The orbits for the outer gaseous planets are vastly different. Jupiter is in a non-nominal state with angular momentum parameters of m = 9/2 and s = −1/4, such that $\mathbf{J}_{orb}{}^2 = (2m + 1)(2m + 2)\,\sigma^2/4 = 110\,\sigma^2/4$, $\mathbf{J}_{fr}{}^2 = s\,(s + 1)\,\sigma^2 =$

$-3\sigma^2/16$, and $\mathbf{\underline{J}}_{oz}{}^2 = [(m + 1/4)(m + 5/4)]\sigma^2 = 437\sigma^2/16$. $\mathbf{\underline{J}}_o$ is the sum of $\mathbf{\underline{J}}_{oz}$ and $\mathbf{\underline{J}}_x$ with a square magnitude of $\mathbf{J}_{oz}{}^2 + \sigma^2 = 453\sigma^2/16$ for $s_A = 1$, and is inclined to $\mathbf{\underline{k}}_o$ at arc-tan$[4/(453)^{1/2}] \cong 10.64$ degrees, as indicated in Figure 6.2. The orbit is aligned with $\mathbf{\underline{J}}_{orb\text{-}x}$, the sum of $\mathbf{\underline{J}}_{orb}$ and $\mathbf{\underline{J}}_x$, with a square magnitude of $\mathbf{\underline{J}}_{orb\text{-}x}{}^2 = \mathbf{\underline{J}}_{orb}{}^2 + \sigma^2 = 114\sigma^2/4$, which exceeds $\mathbf{\underline{J}}_o{}^2$ as depicted in Figure 6.2. The negative value of s indicates a frame rotation component that opposes $\mathbf{\underline{J}}_{orb}$ to produce $\mathbf{\underline{J}}_{oz}$ and $\mathbf{\underline{J}}_o$. The planetary spin fluctuates about $\mathbf{\underline{J}}_{orb\text{-}x}$ with an unperturbed value of $\nu = \gamma = $ arc-tan$(3/456)^{1/2} \cong 4.64$ degrees. The current value of γ is 3.12 degrees.

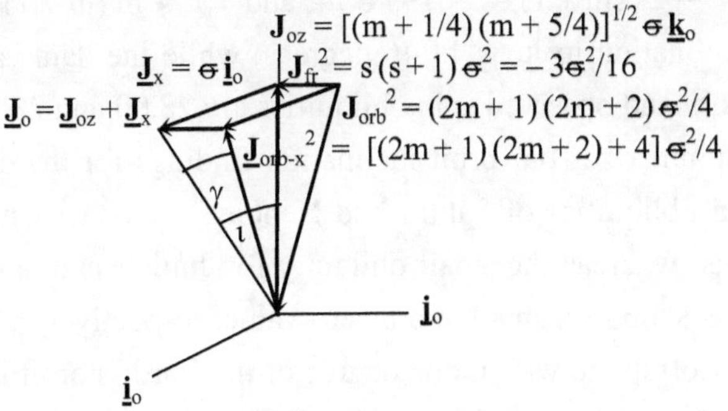

Figure 6.2. Vector Relationships For Jupiter: $m = 9/2$, $s_A = 1$, $s = -1/4$

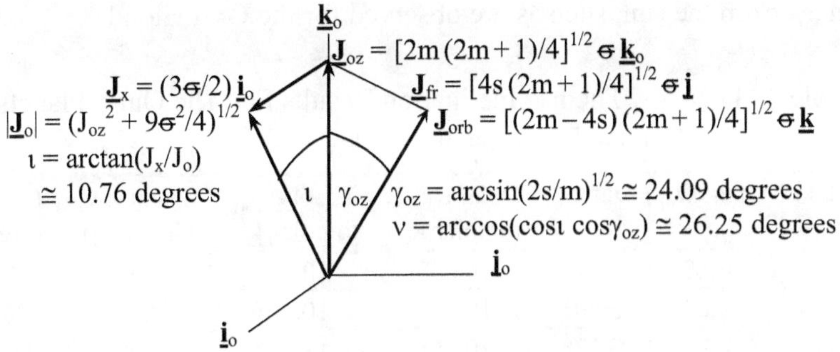

Figure 6.3. Vector Relationships For Saturn: $m = 15/2$, $s_A = 3/2$, $s = 5/8$

The remaining outer planets are sufficiently far from Jupiter that its influence is limited to a perturbation of the solar influence. Saturn's vector diagram in Figure 6.3 is based on $m = 15/2$, $s_A = 3/2$, and $s = 5/8$. Thus, $J_{orb}{}^2$ is $(2m - 4s)(2m + 1)\sigma^2/4 = 50\sigma^2$ and $J_{fr}{}^2 = 4s(2m + 1)\sigma^2/4 = 10\sigma^2$, with $J_{oz}{}^2 = 2m(2m + 1)\sigma^2/4 = 60\sigma^2$. The theory inclination of $\mathbf{\underline{J}}_{orb}$ to $\mathbf{\underline{k}}_o$ is 10.76

degrees, and the data value is 10.72 degrees. The unperturbed spin obliquity of ν is 26.25 degrees, compared to its observed value of 26.73 degrees.

Uranus is in the state m = 19/2, $s_A = 2$, and s = 0, without frame rotation. The theory value of ι is arc-tan$[(2)/(399/4 + 4)^{1/2}]$ = 11.11 degrees. Its retrograde spin inclination of 97.77 degrees lies almost in the plane of its orbit. It is reasonable to attribute its spin and principal moons, also in retrograde orbits in this plane, to remnants of a third body collision similar to the Moon.

The diagram for Neptune is analogous to that for Saturn, and is based on m = 14, $s_A = 3$, and s = 13/8. The theory values are $J_{orb}^2 = (m - 2s)(m + 1)\sigma^2 = 645\,\sigma^2/4$, $J_{fr}^2 = 2s\,(m + 1)\,\sigma^2 = 195\,\sigma^2/4$, and $J_{oz}^2 = m\,(m + 1)\,\sigma^2 = 210\,\sigma^2$. The theory inclination to $\underline{\mathbf{k}}_o$ is 11.46 degrees, while the data value is 11.41 degrees. Theory and observed spin obliquities are 28.00 and 28.32 degrees.

Table 4 summarizes our orbit inclination findings for the outer planets. The large spin obliquities of Saturn and Neptune are consistent with major frame rotations, whereas the small obliquity for Jupiter and spin configuration for Uranus support a small and a zero value, respectively. The fact that all of the outer orbits lie within one degree of the Earth's orbit inclination to the polar axis $\underline{\mathbf{k}}_o$ indicates that the bulk of the primordial solar system lay near this plane. Our theory also affords a virtual continuum of states at large distances from the sun, such as are observed for the Oort Cloud.

Table 4. Data Vs. Theory: Inclination Results For The Outer Planets †
Solar Frame Parameters: $\iota_s = 11.31°$, $\Omega_{s+} = 222°$

Planet	Data: Ω_p	Data: ι_p	Data: ι_{sp} to polar $\underline{\mathbf{k}}_o$	Theory: ι_{sp} to polar $\underline{\mathbf{k}}_o$	Theory: ν (spin avg)	Obser: ν (present)
Jupiter	100.49°	1.31°	10.65°	10.64°	4.65°	3.13°
Saturn	113.64°	2.49°	10.72°	10.76°	26.25°	26.73°
Uranus	74.00°	0.77°	10.66°	11.11°	0.00°	97.77°R
Neptune	131.78°	1.78°	11.41°	11.46°	28.00°	28.32°

Other parameters for the outer planets may also provide a reasonable fit of some of the orbits. However, the results displayed in Tables 2, 3, and 4 should be sufficient to validate the wave theory and the existence of a vector

† Orbital elements for the outer planets vary monthly in the fifth and sixth digits.

potential in the solar gravitational field. Only someone who has never attempted a fitting process would fail to appreciate these results.

An Overview Of Planet-Moon Systems

The fact that the principal moons of Jupiter and Saturn have little or no inclination to their planetary spin axes † indicates that an orbit-level vector potential does not exist in the planetary fields themselves. However, an approximate relationship proportional to $m(m+1)$ between successive orbits for the principal moons of Jupiter and Saturn has previously been noted and used by analysts seeking to attribute lunar surface temperatures to tidal effects. The relationship supports a contention that the wave theory applies to the planet-moon systems. In addition, the scaling of lunar orbit radii in apparent proportion to planetary mass suggests that Mg_o/a_o is a constant independent of the source mass, and we shall proceed under this hypothesis.

Table 5. Data Vs. Theory: Orbit Radii For Principal Moons Of Jupiter
$a_{Jo} = 2.72 \times 10^4$ kilometers

Principal Moon	Body axis dimensions, or diameter (km)	Incl. to Jupiter's equator	Data: mean orbit radius (10^4 km)	Theory: orbit radius unperb (10^4 km)	Theory: m-value
Metis	$60 \times 40 \times 34$	0.06°	12.77	10.3	3/2
Adrastea	$20 \times 16 \times 14$	0.03°	12.87	10.3	3/2
Amalthea	$250 \times 146 \times 126$	0.374°	18.14	16.5	4/2
Thebe	$116 \times 98 \times 84$	1.076°	22.19	24.0	5/2
Io	3660	0.050°	42.17	43.3	7/2
Europa	3121	0.471°	67.10	68.1	9/2
Ganymede	5262	0.204°	107.0	98.3	11/2
Callisto	4821	0.205°	188.3	175.3	15/2

Since a_o for the sun is 2.85×10^7 kilometers and the mass ratio of Jupiter to the sun is about 1/1047, we expect its base lunar parameter to be about a_{oj}

---

† See, Canup, R. and Ward, W., *Origin of Europa and the Galilean Satellites* (2009), University of Arizona Press, Tucson, attributing the formation of "Jupiter's regular satellites from a circumplanetary disk, a ring of accreting gas and solid debris analogous to a protoplanetary disk."

$\cong 2.72 \times 10^4$ kilometers. Table 5 compares the observed mean orbit radii with our theory values for the Jovian moons using only the m (m + 1) relationship. Metis and Adrastea seem to be parts of a common state, and we note that a ring of mass periodically forms below the first moon, but disperses in time. We expect a poor fit of the data since we have not included the proximate effects of the moons' own gravitational fields on each other. However, our overall results should be acceptable to first order. The inclined outer lunar orbits appear to be based on higher order Legendre functions.

Using a mass ratio to the sun of 1/3505 for Saturn, the theory value for a_{so} is 0.82×10^4 kilometers. However, the observed orbit radius of Saturn's first principal moon is 1.81×10^5 kilometers, more than ten times the theory value for m = 1. In fact, the minimum orbit radius and those of several other states afforded by the theory are located inside the mean planetary radius of 5.8×10^4 kilometers. The mean orbit radius of Saturn's largest moon Titan is 1.23×10^6 kilometers, corresponding roughly to m = 12. Our problem is further complicated by the fact that the smaller value of a_{so} equates to closer lunar spacing. These circumstances, including the huge relative mass of Titan, requires a many-body approach to compute the orbit radii, which is beyond the scope of our efforts.

In addition to the gravitational influences of neighboring moons on the orbits of Saturn's satellites, Jupiter's planetary field further affects them. We estimate that Saturn's rings reflect regions of stability smeared by the two gravitational influences. However, the magnitudes of the orbit radii of Saturn's moons qualitatively support our contention that $(Mg_o/a_o)^{1/2}$ is a universal constant approximately equal to 68.3 kilometers per second. We feel confident that the inclusion of third-body Lagrangian terms for the orbits of the moons of Jupiter and Saturn will support our theory.

Considering other moons in the solar system, the two moons of Mars – Phobos and Deimos – are sufficiently close to the planetary surface to prevent their orbits from becoming unstable due to Jupiter's perturbing influence. However, they hardly provide a test for the spacing predicted by our theory. As for the Earth, the size of our Moon in relation to the planet is anomalous among planet to moon relationships in the solar system, and we have treated the Moon and the Earth as forming a binary planetary system.

7 – SPIN OBLIQUITY, PRECESSION, NUTATION, AND ICE AGES

Although the Earth's spin axis generally points toward the North Star, it is observed to nutate (nod) in an irregular elliptic pattern around its mean pointing direction by 10 meters or so over an average period of about 420 days. The nutation is called the *Chandler Wobble* after Seth Carlo Chandler, its discoverer in 1891. The average spin inclination (*obliquity*) to the Earth's orbit is now 23.45 degrees, but it is decreasing by about 1.35 ten-thousandths of a degree per year. Astronomers estimate that it will reach a minimum of 22 degrees some 10,000 years in the future, before reversing and proceeding to an opposite extreme of about 25 degrees. Simultaneously, the intersection of the Earth's equator with its orbit at the equinoxes is drifting steadily westward over a period of about 26,000 years, *i.e., precession of the equinoxes.* These phenomena are small frame rotations due to gravitational torques.

During the 1700s Leonhard Euler modeled the Earth's body without orbital constraints as a spinning top whose spin axis undergoes undriven harmonic nutation over a period of $2\pi\, I_s/(I_s - I_a)/\check{s} \cong 306$ days, where I_s is the body moment of inertia about its spin axis, I_a is the moment about any axis in the Earth's equatorial plane, $\check{s}$ is 2π radians per day, and $(I_s - I_a)/I_s$ is approximately 0.003273. Euler's computation offers no explanation for the nutation amplitude or the precession of the equinoxes, and the period provided by his data-limited approach errs in the wrong direction relative to $\check{s}$.

Recent analyses apply Euler's dynamical equation to body moments for an oblate planet to estimate a small frame rotation ϕ' produced by solar and lunar torques on the inclined spin axis. These efforts successfully model the Earth's equinox precession, † but fail to address spin axis nutation. Proper modeling of the torques explains the wobble and secular nutation, in addition to equinox precession. However, the analysis requires some in-depth efforts.

---

† See, Goldstein, H., *Classical Mechanics*, 1980 ed., *supra,* pp. 225-232; see also, Becker, R. A., *Introduction to Theoretical Mechanics*, *supra*, pp. 292-296. See further, *Excitation of the Chandler Wobble*, Gross, R., JPL, Pasadena, CA (1999), attributing the wobble to "atmospheric and oceanic processes", based on correlations between the wobble and the cited processes, and incorrectly assigning the cause versus the effect.

Figure 7.1 depicts the Earth's angular momentum vectors at the orbit-level in an $\underline{\mathbf{i}}_o\underline{\mathbf{j}}_o\underline{\mathbf{k}}_o$ inertial frame, where ξ' represents principal frame rotation. The body-based vector $\mu\underline{\mathbf{J}}_k$ lies along a $\underline{\mathbf{k}}$ axis inclined to $\underline{\mathbf{k}}_o$ in the $\underline{\mathbf{j}}_o\underline{\mathbf{k}}_o$ plane at an angle γ. Its magnitude is $I_z W_o$, where μ is the planetary mass, I_z is the orbital moment of inertia, and W_o is its orbital angular frequency at a special angle γ_o. A torque due to the curl of the vector potential $\underline{\mathbf{A}}_g$ causes frame rotation of ξ', resulting in $\mu\underline{\mathbf{J}}_{oz} = I_z(W_o + \xi'_o \cos\gamma_o)\,\underline{\mathbf{k}}_o$ and $\mu\underline{\mathbf{J}}_{oy} = I_z\xi'_o \sin\gamma_o\,\underline{\mathbf{j}}_o$. The sum vector $\underline{\mathbf{J}}_{sum} = \underline{\mathbf{J}}_{oy} + \underline{\mathbf{J}}_{oz}$ reduces to $\underline{\mathbf{J}}_{oz}$ at a critical angle γ_{oz}, as $\underline{\mathbf{J}}_{oy}$ vanishes and $\xi'_o \cos\gamma_o$ is aliased into the expression of W_{oz}. $\underline{\mathbf{A}}_g$ also creates $\underline{\mathbf{J}}_x = \sigma\,\underline{\mathbf{i}}_o/2$ for the Earth, which adds to $\underline{\mathbf{J}}_{oz}$ to produce the observed orbit vector $\mu\underline{\mathbf{J}}_o$ inclined at an angle ι along the $\underline{\mathbf{i}}_o$ axis.

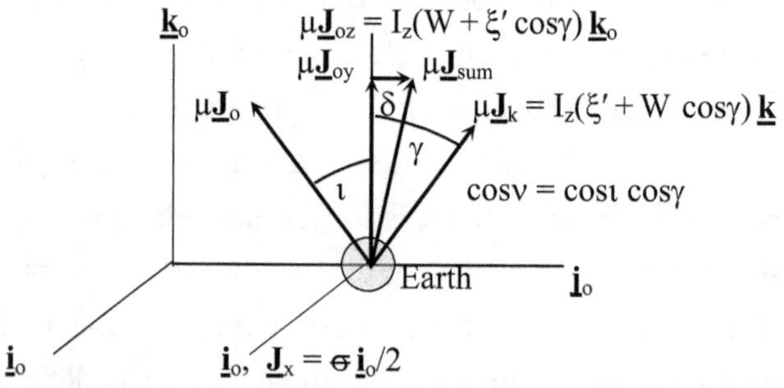

Figure 7.1. Orbit-Level Frame Relationships For The Earth

If we align the body $\underline{\mathbf{k}}$ axis with the planetary spin $\underline{\mathbf{s}}$, the angle γ will vary slowly with the spin direction. Changes to W and ξ' in the $\underline{\mathbf{ijk}}$ body frame caused by variations in γ have been specified above. For the Earth we found the value of γ in inertial space to be $\gamma_{oz} \cong 20.706$ degrees, at which W becomes 1.0690 W_o as ξ' and the small angle δ in the body frame of Figure 7.1 become zero. It is only at γ_{oz} that $\underline{\mathbf{J}}_k = I_z(\xi' + W\cos\gamma)\,\underline{\mathbf{k}}$ is actually $\underline{\mathbf{J}}_{orb}$, even though its magnitude remains constant at $|\underline{\mathbf{J}}_{orb}|$. Let us further use the $\underline{\mathbf{ijk}}$ frame to include precession of ϕ' for the Earth's body about the $\underline{\mathbf{k}}_o$ axis and nutation of γ' about the $\underline{\mathbf{i}}$ axis, as indicated in Figure 7.2. The $\underline{\mathbf{i}}$ axis is initially aligned with $\underline{\mathbf{i}}_o$, but it rotates with ϕ' about $\underline{\mathbf{k}}_o$, and the body $\underline{\mathbf{j}}$ axis is specified by $\underline{\mathbf{k}} \times \underline{\mathbf{i}}$ in the rotating and nutating frame.

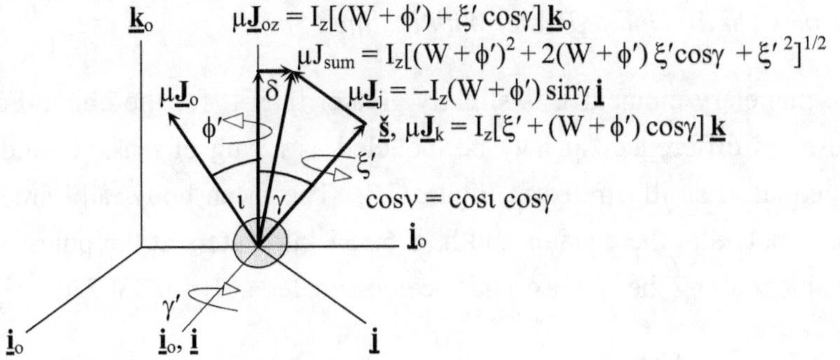

Figure 7.2. Frame Rotations In The Body Frame Of Reference

The form $\mu J_{sum} = I_z[(W+\phi')^2+2(W+\phi')\,\xi'\cos\gamma+\xi'^2]^{1/2}$ is the magnitude of the sum of $\mu\mathbf{\underline{J}}_j = -I_z(W+\phi')\sin\gamma\,\mathbf{\underline{i}}$ and $\mu\mathbf{\underline{J}}_k = I_z[\xi'+(W+\phi')\cos\gamma]\,\mathbf{\underline{k}}$ in the $\mathbf{\underline{ijk}}$ frame. The magnitude of $\mu\mathbf{\underline{J}}_k$ is I_zW_o, as $\mu\mathbf{\underline{J}}_{oz} = I_z(W+\xi'\cos\gamma+\phi')\,\mathbf{\underline{k}}_o$ remains constant in magnitude and direction. Frame rotation of γ' about $\mathbf{\underline{i}}$ is multiplied by the orbital moment $I_x = I_z/2$, as for a ring of mass, and by I_a, the body moment about the $\mathbf{\underline{i}}$ axis. Frame rotation of ϕ' about $\mathbf{\underline{k}}_o$ can be resolved into $\phi'\cos\gamma\,\mathbf{\underline{k}}$ and $-\phi'\sin\gamma\,\mathbf{\underline{i}}$, where the latter is multiplied by the body moment I_a. I_s multiplies angular frequencies of $\phi'\cos\gamma$ and ξ', and the spin š.

$$E = I_z\gamma'^2/4 + I_z[(W+\xi'\cos\gamma+\phi')^2 + (\xi'\sin\gamma)^2]/2 + V_o + \mu Q$$
$$+ I_a(\gamma'^2+\phi'^2\sin^2\gamma)/2 + I_s(\check{s}+\xi'+\phi'\cos\gamma)^2/2 + V_{b1} + V_{b2}. \qquad (1)$$

We now include the effects of small frame rotations of ϕ' and γ' in the total energy, expression 1, where we are suppressing separable terms produced by the vector potential $\mathbf{\underline{A}}_g$ and are approximating the orbits as being circular. The inclined spin configuration produces the two small potential terms, V_{b1} and V_{b2}, whose partial derivatives with respect to ϕ and γ model torques on the Earth's spinning body. These torques create the small frame rotations of ϕ' and γ', which complement $\xi'\,\mathbf{\underline{k}}$ to form a complete set of Euler rotations. The ξ' rotation about $\mathbf{\underline{k}}$ does not affect the body motion, but the frame rotations of ϕ' and γ' appear as precession and nutation for the Earth's body. However, all three frame rotations are offset by angular momentum components in inertial space and do not produce any movement at the orbit level. We cannot overemphasize this fact, since it underlies orbit stability.

Torques On An Oblate Planet With Inclined Spin

The planetary moment I_s is slightly greater than I_a for the oblate Earth, and the mass difference $\Delta\mu$ may be modeled as a ring of mass around the Earth's equator, as illustrated in Figure 7.3. The mean body radii are $h_a \cong$ 6,378 kilometers at the equator and $h_s \cong$ 6,357 kilometers at the poles. The body moment along the spin axis has been estimated at $I_s = 0.3307\ \mu h_a^2$. †

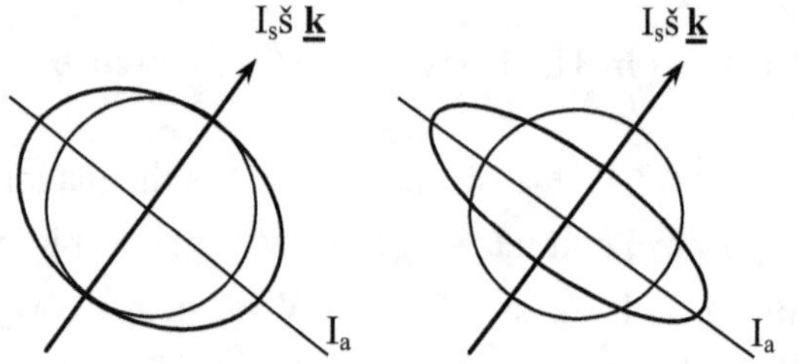

Figure 7.3. Oblateness Modeled As An Equatorial Ring Of Mass

The difference between I_s and I_a has been computed from satellite tracking data †† using the term $J_2 = (I_s - I_a)/(\mu h_a^2) = 1082.6 \times 10^{-6}$ as it appears in $V_{02}(r_e) = (\mu g_o/r_e)\,[1 - J_2(h_a/r_e)^2\,(3\sin^2\lambda - 1)/2]$, where $V_{02}(r_e)$ is the second order geopotential, r_e is the distance from the planet's center, and λ is the latitude on the Earth. The accepted value of $(I_s - I_a)/I_s$ is about 0.003273.

The locations on the spin-stabilized ring, perpendicular to **k** in Figure 7.3, are subject to small solar force differences at all times during the orbit except the equinoxes. The forces at the edges of the ring during a winter

---

† Concentrations of dense elements in the Earth's interior produce its principal moment of $I_s = 0.3307\ \mu h_a^2$, rather than a uniform sphere value of $0.4\ \mu h_a^2$. See, *e.g.*, Hancock, P., and Skinner, B., "Moment of Inertia and Precession", *The Oxford Companion to the Earth 2000*; http://www.encyclopedia.com, 5 July 2012. An acceptable mass model is composed of uniform layers ending at distances of 1220, 3480, 6350, and 6378 kilometers from the Earth's center, with respective average layer densities of 13.36, 10.90, 4.48, and 2.35 grams per cubic centimeter. See, *e.g.*, Beatty, Petersen, and Chaiken, (eds.), *The New Solar System*, Sky Pub. Corp., Cambridge, MA (1999), pp.113-115.

†† See, Bate, Mueller, and White, *Fundamentals of Astrodynamics*, *supra*, at p. 422.

solstice are indicated by $\underline{\mathbf{F}}_+$ and $\underline{\mathbf{F}}_-$ in Figure 7.4, as compared to $\underline{\mathbf{F}}$ at $\underline{\mathbf{r}}$ at this same time. The non-central force components $\underline{\mathbf{F}}_{nc+}$ and $\underline{\mathbf{F}}_{nc-}$ produce a torque of $\underline{\mathbf{r}} \times \underline{\mathbf{F}}_{nc}$ along $\underline{\mathbf{i}}_o$ which does not average to zero over the orbit.

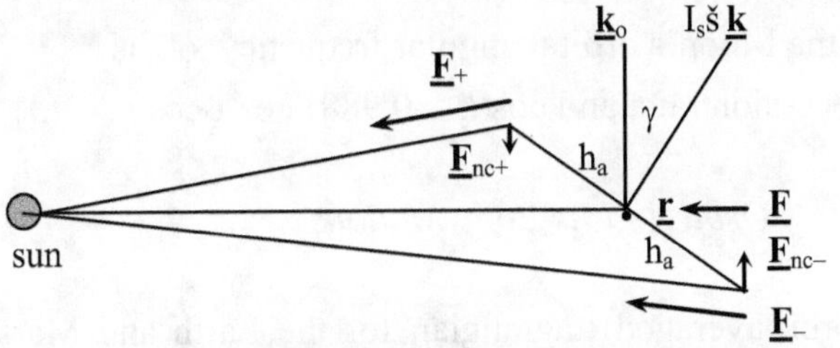

Figure 7.4. Solar Force On An Equatorial Ring

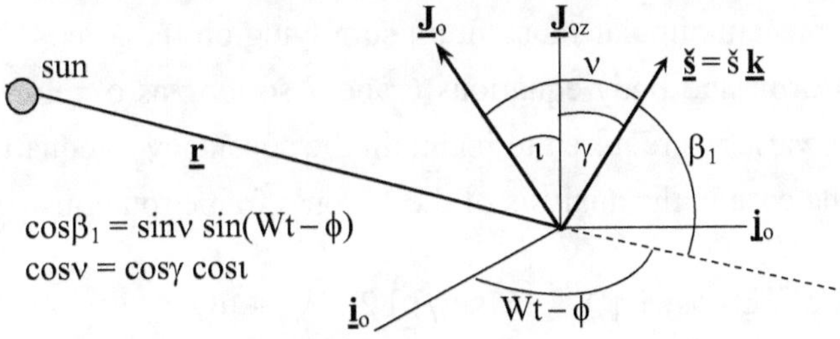

Figure 7.5. Body Angles For Modeling The Solar Torque

We have derived the form of the solar torque produced by stabilized spin in Appendix 2, and shown that it can be specified by the derivative of a quadrupole energy term such as expression 2. V_{b1} depends on the angle β_1 depicted in Figure 7.5 which is a function of γ, ι, and $Wt - \phi$. The vector potential $\underline{\mathbf{A}}_g$ has tilted the orbit vector away from $\underline{\mathbf{k}}_o$ at a constant angle ι and increased the obliquity angle from γ to ν. It is convenient to use Kepler's third law, $W^2 = g_o M/r^3$, to compute the effects of V_{b1}.

$$V_{b1} = -(I_s - I_a)(Mg_o/r^3)(3\cos^2\beta_1 - 1)/2,$$
$$\text{where } \cos\beta_1 = \sin\nu \sin(Wt - \phi). \qquad (2)$$

The Moon's motion also creates a torque the Earth's spin axis specified by the partial derivative of V_{b2}, expression 3, with respect to γ. V_{b2} is derived in Appendix 7, where Figure A7.1 depicts the involved angles and their derivatives. We may use the present value of W = 2π radians per year to calculate V_{b2}, but it is not an explicit function of W nor of ϕ'.

$$V_{b2} = -2.20\,W^2 (I_s - I_a)\,[3\sin^2\zeta \sin^2(w_m t - \phi'_m t) - 1)]/2, \quad \text{where}$$

w_m is the Moon's orbital angular frequency, ϕ'_m is its precession rate, and $\cos^2\zeta \cong 0.9881 \cos^2\iota \cos^2\gamma$. (3)

Constants Of The Earth's Orbit And Its Body Motions

Expression 4 is the orbit-averaged Lagrangian for the Earth and Mars (without V_{b2} for Mars), including the body equations. Taking its partial with respect to ξ', we obtain equation 5. Since there is no torque along **k**, the result represents a constant angular momentum sum lying on the **k** axis that can be separated as orbit and body equations 6 and 7 so long as ϕ' remains constant. The spin š varies as required to maintain the constancy of equation 7, as we saw to be the case in the analysis of the Moon's frame rotations.

$$L = I_z\gamma'^2/4 + I_z[(W + \xi' \cos\gamma + \phi')^2 + (\xi' \sin\gamma)^2]/2 - V_o - \mu Q$$
$$+ I_a(\gamma'^2 + \phi'^2 \sin^2\gamma)/2 + I_s(\check{s} + \xi' + \phi' \cos\gamma)^2/2 - V_{b1} - V_{b2}, \quad (4)$$

$$d/dt\,\{I_z[(W + \phi')\cos\gamma + \xi'] + I_s(\check{s} + \xi' + \phi' \cos\gamma)\} = 0, \quad (5)$$

$$I_z[(W + \phi')\cos\gamma + \xi'] = \mu J_k = I_z W_o, \quad \text{a constant}, \quad (6)$$

$$I_s(\check{s} + \xi' + \phi' \cos\gamma) = I_s \check{s}_o, \quad \text{a constant}. \quad (7)$$

Next equating d/dt ($\partial L/\partial\phi'$) to $\partial L/\partial\phi$, where $\partial V_{b2}/\partial\phi$ is zero, we obtain equation 8, where the sum in large brackets is the angular momentum lying on the $\mathbf{k}_o$ axis. This sum is also constant except for a small oscillation due to sin2(Wt – φ), which averages to zero over Wt, expression 9. We thus identify $I_z[(W + \phi') + \xi' \cos\gamma]$ as the magnitude of $\mu\mathbf{J}_{oz}$ along $\mathbf{k}_o$, expression 10. The vector $\mu\mathbf{J}_j = -I_z(W + \phi') \sin\gamma\, \mathbf{i}$ adds to $\mu\mathbf{J}_k = I_z[(W + \phi') \cos\gamma + \xi']\,\mathbf{k}$ to form $\mu\mathbf{J}_{sum}$. Consistent with observations, we will treat ϕ' as a negative constant $-\phi_o'$, as W and ξ' continue to vary with γ in the body frame, expression 11, with J_k and J_{oz} being constant and ϕ' being negligible at the orbit-level.

$$d/dt\,\{I_z[(W+\phi')+\xi'\cos\gamma]+I_s(\check{s}+\xi'+\phi'\cos\gamma)\cos\gamma+I_a\,\phi'\sin^2\gamma]\}$$
$$= -\partial V_{b1}/\partial\phi = 3(I_s - I_a)\,(Mg_o/r^3)\sin^2\nu\,\sin(Wt-\phi)\cos(Wt-\phi), \quad (8)$$

$$-I_s\check{s}_o\,\gamma'\sin\gamma \cong 3\,(I_s-I_a)\,(Mg_o/r^3)\sin^2\nu\,\sin(Wt-\phi)\cos(Wt-\phi), \quad (9)$$

$$I_z[(W+\phi')+\xi'\cos\gamma] = \mu J_{oz}, \qquad \text{a constant}, \quad (10)$$

$$W = W_o\,(J_{oz}/J_{orb} - \cos\gamma)/\sin^2\gamma, \qquad \xi' = W_o - W\cos\gamma. \quad (11)$$

The partials of the Lagrangian with respect to γ' and γ provide equation 12. The term Q is not a function of ϕ or ϕ'. Next setting $I_z(W+\phi')\,\xi'\sin\gamma = -\mu\,\partial Q/\partial\gamma$ at the orbit-level, we obtain expression 13 as the constant body energy E_b from which all of the body motions are derived.

$$(I_z/2+I_a)\,\gamma'' = -I_z(W+\phi')\,\xi'\sin\gamma - \mu\,\partial Q/\partial\gamma - I_s\check{s}_o\,(\phi'\sin\gamma)$$
$$+I_a(\phi'\cos\gamma)\,(\phi'\sin\gamma) - \partial V_{b1}/\partial\gamma - \partial V_{b2}/\partial\gamma, \quad (12)$$

$$E_b = (I_z/2 + I_a)\,\gamma'^2/2 + I_a(\phi'^2\sin^2\gamma)/2 + V_{b1} + V_{b2},$$
$$+ I_s(\check{s} + \xi' + \phi'\cos\gamma)^2/2, \qquad \text{constant body energy.} \quad (13)$$

Precession Of The Earth's Equinoxes

Using $I_z(W+\phi')\,\xi'\sin\gamma = -\mu\,\partial Q/\partial\gamma$ and ignoring $I_a\,\phi'^2\sin\gamma\cos\gamma$ in equation 12, we obtain equation 14 after averaging the partials of V_{b1} and V_{b2} over time. The usual procedure of equating the sum on the right side of the equation to zero applies only when γ'' is zero. Fortunately, we can compute the critical value of γ where $\gamma'' = 0$ by inputting the observed values of ϕ' and other terms. Our efforts are aided by defining the parameter $\epsilon^2 = (I_s - I_a)/I_z$ and using $W_e = 2\pi$ radians per year as a reference value to compute $\partial V_{b2}/\partial\gamma$.

$$I_z\gamma''/2 = -I_s\check{s}_o\,\phi'\sin\gamma + (3/2)\,(I_s-I_a)\,W^2\cos^2\iota\,\sin\gamma\cos\gamma$$
$$+ (3/2)\,(I_s-I_a)\,(w_m^2/81.3)\,(\sin\zeta\cos\zeta)\,\partial\zeta/\partial\gamma. \quad (14)$$

Setting $I_s - I_a = J_2\,\mu h_a^2 = (1.0826\times10^{-3})\,\mu h_a^2$ with $h_a = 6{,}378$ kilometers and using $I_z = \mu a_n^2$ for the Earth's mean orbit radius of $a_n = 1.496\times10^8$ kilometers, we find that $\epsilon^2 = (I_s - I_a)/I_z \cong 1.969\times10^{-12}$ and $\epsilon \cong 1.403\times10^{-6}$.

Turning to the Earth's precession, its observed rate is 50.26 arc-seconds per year. However, 1.83 arc-seconds have been attributed to the gravity fields of other planets, leaving a luni-solar component of $\phi'_{obs} \cong -48.43$ arc-seconds per year $\cong -26.63 \in W_e$, expression 15. Multiplying ϕ'_{obs} by $I_s\check{s}_o/I_z \cong 0.1567 \in W_e$ for $\check{s}_o = 2\pi$ radians per day and $I_s = 0.3307\ \mu h_a^2$, we obtain the second line of the expression. Inputting this form to equation 14 results in equation 16. Labeling the constant ϕ' as ϕ'_{obs}, setting $w_m \cong 13.37\ W_e$, † and using equation 11 to compute W as a function of γ, we find by iterations that γ'' becomes zero at a critical angle of $\gamma_c \cong 21.45$ degrees, expression 17.

$$\phi'_{obs} \cong -2\pi/(26{,}760 \text{ years}) \cong -26.63 \in W_e, \quad \text{for } \in \cong 1.403 \times 10^{-6},$$
$$I_s\check{s}_o\, \phi'_{obs}/I_z \cong -4.172 \in^2 W_e^2, \quad \text{for } W_e = 2\pi \text{ radians per year,} \quad (15)$$

$$\gamma''/(2\in^2 W_e^2) = -4.172 \sin\gamma + [1.5\,(W/W_e)^2 + 3.259] \cos^2\iota \sin\gamma \cos\gamma, \quad (16)$$

$$I_s\check{s}_o\, \phi'_{obs}/I_z \cong -4.482 \in^2 W_e^2 \cos\gamma_c, \text{ where } \gamma'' = 0 \text{ at } \gamma_c \cong 21.45 \text{ degrees.} \quad (17)$$

Using obliquity of ν rather than γ, physics texts usually equate $I_s\check{s}_o\, \phi' \sin\gamma$ to $-\partial V_{b1}/\partial\gamma$ and later include the lunar effects. However, $\partial V_{b2}/\partial\gamma$ must be included at the outset in order to specify γ'' as γ varies, and it is error to ignore $I_z\gamma''/2$ since I_z is so large, even though γ'' is very small. The relationship between $I_z\gamma'^2/2$ and $I_a\phi'^2 \sin^2\gamma$ is comparable to the classical expression for the nutation of a spinning top, except for the magnitudes of $I_z/2$ and I_a.

Far Term Nutation

Having identified the constant body energy E_b as expression 13, independent of orbit-level energy, we calculate E_b by using the current values of ν' and other parameters. Inputting V_{b1} and V_{b2} without averaging over Wt or $w_m t$, and ignoring the small term $I_a(\phi'^2 \sin^2\gamma)/2$, we may rewrite equation 13 as equation 18. Substituting $-4.172 \in^2 W_e^2$ for $I_s\check{s}_o\, \phi'_{obs}/I_z$, expressing ζ as a function of γ, and setting $Wt - \phi \cong Wt$, expression 19 defines an energy

† This is not quite true since w_m and W_e may vary independently in our model, and the lunar orbit radius slowly increases due to tidal friction.

expression E_γ as the sum of terms in equation 13 that are not time-explicit. E_γ is almost constant during any orbital period since γ varies so slowly.

$$\gamma'^2/(4\epsilon^2 W_e^2) = E_\gamma/(I_z\epsilon^2 W_e^2) + 1.5\,(W/W_e)^2 \sin^2\nu \sin^2 Wt + 3.298 \sin^2\zeta \sin^2 w_m t, \qquad (18)$$

$$E_\gamma/(I_z\epsilon^2 W_e^2) = E_b/(I_z\epsilon^2 W_e^2) - 1.0994 + 4.172 \cos\gamma - 0.5\,(W/W_e)^2. \qquad (19)$$

Furthermore, E_γ must be a non-negative function of γ in order for γ'^2 to remain real in equation 18.† But E_γ will become zero at certain values of γ, leaving γ'^2 as the sum of squares of two sinusoidal functions. At such values of γ the two oscillatory functions both become zero at some time during their cycles, and γ'^2 momentarily becomes zero, allowing γ' to reverse its sign.

Let us suppose that γ' is negative after passing through γ_c, so that $\cos\gamma$ is increasing. As γ continues to decrease, the magnitude of the term $0.5\,W/W_e$ in expression 19 grows much faster than $4.172 \cos\gamma$ due to the factor of $\sin^2\gamma$ in the denominator for W in expression 11. Thus, E_γ eventually becomes zero, where γ' reverses sign and becomes positive. Thereafter, $4.172 \cos\gamma$ initially dominates as γ increases, with E_γ retracing its previous steps. But γ' remains positive as $4.172 \cos\gamma$ continues to fall. However, $-0.5\,(W/W_e)^2$ finally overtakes it, with γ' becoming zero at an increased value of γ. Thus, γ progresses from a minimum of γ_{min} where γ'^2 is zero to a maximum of γ_{max}, where γ'^2 again becomes zero and the sign of γ' reverses. Throughout the cycle γ'^2 is greater than zero except at γ_{min} and γ_{max}, and it reaches its maximum absolute value when γ'' is zero. Figure 7.6 is a generic sketch for γ'^2.

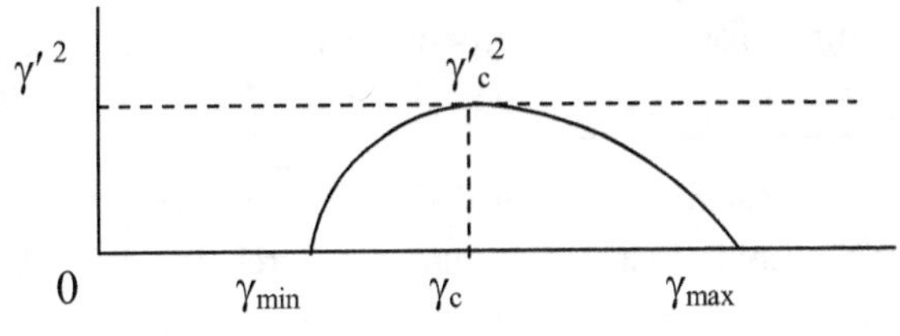

Figure 7.6. Spin Axis Nutation In The Far Term

---

† If we had included the average of $\sin^2 Wt = (1 - \cos 2Wt)/2$ as a part of E_γ/I_z in equation 11, γ'^2 would become negative at $Wt = w_m t = 0$ for small values of E_γ/I_z.

In 1908 a monument was erected in Taiwan to mark the location of the Tropic of Cancer, which has shifted to the south at an average rate of $ds/dt \cong$ 15 meters annually. Using a polar radius of $h_s \cong$ 6,357 kilometers, we find that $d\nu/dt = (ds/dt)/h_s = -2.36 \times 10^{-6}$ radian per year at $\nu_e = 23.45$ degrees. Taking the time derivative of $\cos\nu = \cos\iota \cos\gamma$ with ι held constant and computing W/W_e, expression 20 provides the value of γ'_e at $\gamma_e = 20.68$ degrees.

$$\gamma'_e = -0.5593 \text{ arc-second per year} = -0.3076 \in W_e, \qquad (20)$$

$$E_b = -2.2801\ I_z \in^2 W_e^2. \qquad (21)$$

Setting $E_\gamma/I_z = \gamma'^2_e/4 = 0.0237 \in^2 W_e^2$ at $\gamma = \gamma_e$, we obtain expression 21 for E_b. Knowing E_b in equation 19, and using equation 11 to provide W/W_e as an input, we can now compute numerical values of E_γ as γ varies. We find that $E_\gamma = 0$ at $\gamma_{min} \cong 19.75$ degrees and at $\gamma_{max} \cong 27.65$ degrees. Setting $\cos\nu = \cos\iota \cos\gamma$, we obtain ν_{min} and ν_{max} for the Earth, expression 22. Our estimate of ν_{min} is slightly greater than the largest of astronomers' published estimates, but our value for ν_{max} is considerably greater, resulting in an obliquity range more than twice their estimates † and a much longer period.

Earth's Obliquity Limits: $\quad 22.65 \text{ degrees} \le \nu \le 29.70 \text{ degrees}. \qquad (22)$

Time Of Passage And The Ice Ages

Equating the secular value of $\gamma'^2/4$ to E_γ/I_z in equation 18 and taking its square root, we can use the integral of dt in equation 23 to compute the *time of passage* t_{pa} required for the spin axis to tilt from γ_{min} to γ_{max}. We estimated

---

† See, Dawicki, S., *Changes in Earth's Tilt Control When Glacial Cycles End*, Woods Hole Oceanographic Institute, http://www.eurekalert.org/pub_releases/2005-03/whoicie 032905.php, estimating the limits of ν at $\nu_e \pm 1$ degree. See also, Beatty, et al., *The New Solar System*, *supra,* p. 189, estimating ν at $\nu_e \pm 1.5$ degrees. The estimates are based on observations of ν'_e and the assumptions that ν' is nearly constant and changes sign after a period of 41,000 years. The period being utilized is provided by the difference between the 26,000 year luni-solar precession period and a 71,000 year value attributed to perturbing fields of other planets.

the result by numerically integrating in varying steps of 0.25 to 1.0 degree, using interval midpoints, and doubling t_{pa} to obtain the cycle period. We note that the use of midpoints multiplied by the interval keeps the sum finite even though the integrand denominator becomes zero at the extremes of γ.

$$2\in W_e \int dt = \int d\gamma/[E_b/(I_z\in^2 W_e^2) + 4.172\cos\gamma - 0.5\,(W/W_e)^2 - 1.0994]^{1/2},$$

$$2t_{pa} \cong 105{,}000 \text{ years}, \qquad \text{for } 19.75 \le \gamma \le 27.65 \text{ degrees.} \quad (23)$$

During the first half of the Twentieth Century, Milutin Milanković (1879-1958) modeled the Earth's insolation at the poles as a function of the incidence angles of the sun's rays, and proposed a correlation between the spin obliquity and the Earth's glacial cycles. But it was not until the 1970s that researchers were able to test the theory based on the compositions of ice core samples collected in Greenland and Antarctica. The samples show that the durations of the Earth's last four glacial cycles have been about 125,000, 110,000, 90,000, and 95,000 years, in order from the present and proceeding backward in time. † By incorporating various orbit effects, climatologists have attempted to explain the glacial cycles as multiples of the 41,000 year period proposed by astronomers, but our result provides a more credible validation of the Milanković theory. Using a base period of 105,000 years, the variability in ice age durations may be attributed to the varying eccentricity of the Earth's orbit and other phenomena. ††

The Chandler Wobble

The Chandler Wobble is a small oscillation about the Earth's median spin direction over an average period of about 420 days. The deviation

---

† See, Weart, S., *The Discovery of Global Warming*, Harvard Univ. Press (2008, 2011); see also, American Institute of Physics, http://www.aip.org/history/climate/cycles.htm, citing Petit et al., *Nature* **399**: 429-436 (1999), and discussing temperature estimates based on the relative abundances of O^{18} and O^{16} oxygen isotopes.

†† There is also dependence on volcanic activity and un-modeled solar phenomena, as well as the variability of other orbit parameters, such as are observed for the Moon.

roughly traces out an ellipse on the Earth's surface perpendicular to the polar radius h_s. † Let us formulate the wobble as $h_s\,\Delta\gamma = h_s \int (d\gamma/dt)\,dt$, where h_s = 6,357 kilometers and $\Delta\gamma$ is the integral of $d\gamma/dt$ provided by the time-explicit components of V_{b1} and V_{b2} over a specified period of time. During an average lunar orbit cycle of 27.3 days the solar component V_{b1} is nearly constant and $\Delta\gamma$ is driven by $w_m t$. But the lunar effect averages almost to zero over its successive periods, and the solar torque dominates the wobble for longer periods. Ignoring the lunar oscillation term and taking the square root of equation 18, we obtain equation 24 for the temporal behavior of γ'. Integration over a quarter year period at $W = W_e$ leads to an elliptical integral of the second kind, equation 25, where χ is the complement of $W_e t$ and the oscillation amplitude is of the order $\in = 1.403 \times 10^{-6}$ radian, as observed.

$$\gamma' = -2\in W_e\, b_\gamma\,[1 - (1.5 \sin^2 v_e)(\cos^2 W_e t)/b_\gamma^{\,2}]^{1/2},$$
$$\text{where } b_\gamma^{\,2} = E_\gamma/(I_z \in^2 W_e^{\,2}) + 1.5 \sin^2 v_e, \qquad (24)$$

$$\Delta\gamma = 2\in b_\gamma \int (1 - k^2 \sin^2\chi)^{1/2}\, d\chi, \quad \text{where } \chi = \pi/2 - W_e t,$$
$$0 \le W_e t \le \pi/2,\ k^2 = 1.5\,(\sin^2 v_e)/b_\gamma^{\,2}, \text{ and}$$
$$b_\gamma^{\,2} = (0.3076)^2/4 + 1.5 \sin^2 v_e, \text{ at present,} \qquad (25)$$

$$\Delta\gamma_1 = 2\in b_1 W \int (\sin Wt)\, dt, \qquad \text{when } E_\gamma = 0, \text{ and } b_1 = (1.5)^{1/2} \sin v,$$
$$= 2\in b_1 (1 - \cos Wt), \qquad \text{for } 0 \le W_e t \le \pi/2. \qquad (26)$$

Whenever E_γ is zero, b_γ in equation 24 will be $b_1 = (1.5)^{1/2} \sin v$ and $\Delta\gamma_1$ will be given by equation 26 over the period $\Delta t_1 = (\pi/2)/W_e$. The term $2\in b_1$ is the oscillation amplitude of $\Delta\gamma_1$ during Δt_1, and the displacement would be $2\in b_1$ at the end of the 91.3 day period. Although the integrand still represents an oscillation when E_γ is not zero, the period and the amplitude are intertwined, and the period becomes greater than 91.3 days, similar to a

† Although our analysis models the torque as confined to the **i** axis, the wobble pattern over any given year is somewhat more complicated due to the varying location of the Moon and the orientation of its orbit vector at the time of observation. In addition, we have not estimated a time lag or dampening of the motion due to more complex modeling of the Earth's inertial moment.

pendulum of arbitrary amplitude. [†] If E_γ were zero at the present time for ν_e = 23.45 degrees, b_γ would be $b_l \cong 0.4874$. But it is in fact $b_e \cong 0.5111$, and we obtain $k_e \cong 0.9536$ by using $E_\gamma/I_z \cong 0.0237 \in^2 W_e^2$ for the Earth. The value of the equation 25 integral is then 1.0969, [††] so that Δt_e in equation 27 is about 1.150 Δt_l for Δt_l = 91.3 days. Thus, the wobble period should be 1.150×365.25 days $\cong$ 420 days. However, the Moon's orbit period does not divide evenly into 420 days, and its motion causes the observed period to vary somewhat during any given year.

$$4\,\Delta t_e \cong (b_e/b_l)\,(1.0969)\,(4\,\Delta t_l) \cong 1.150 \times 365.25 \text{ days} \cong 420 \text{ days,}$$

$$\text{estimated average period of the Chandler wobble.} \qquad (27)$$

Precession And Nutation As Angular Momentum Components

It may seem anomalous that frame rotations of ϕ' and γ' are observed in the motion of the planet body, whereas they appear at the orbit level simply as components of constant angular momentum. However, this is the reality. In a dispute with NASA, the Sirius Research Group has shown conclusively that the transits of Venus are not correctly modeled when equinox precession is included in the inertial reference frame. [†††] Our analysis treats both $I_z\gamma'/2$ about $\underline{\mathbf{i}}$ and $I_z\phi'$ about $\underline{\mathbf{k}}_o$ as small angular momentum terms in the $\underline{\mathbf{i}}_o\underline{\mathbf{j}}_o\underline{\mathbf{k}}_o$ reference frame, where γ is fixed at γ_{oz} and $\mu\underline{\mathbf{J}}_o$ remains constant in magnitude and direction, absent orbit-level perturbations. Thus, neither precession of ϕ' nor nutation of γ' in the $\underline{\mathbf{ijk}}$ body frame changes the inclination γ of the orbital angular momentum vector $\mu\underline{\mathbf{J}}_{orb}$ to the $\underline{\mathbf{k}}_o$ solar axis; neither do they change the angle ν between $\mu\underline{\mathbf{J}}_{orb}$ and the orbit vector $\mu\underline{\mathbf{J}}_o$.

[†] See, Becker, R. A., *Introduction to Theoretical Mechanics*, *supra*, pp. 267-269.

[††] See, Selby, S. (ed.), *CRC Standard Mathematical Tables*, 15th Ed., The Chemical Rubber Co., Cleveland, OH (1967), pp. 479-481.

[†††] See, Homann, U., "Transits of Venus vs. NASA's Astronomical Data", 21 April 2004, http://www.siriusresearchgroup.com.articles/Nasa-Venus-Transits.html. In an inimitable way, status quo authorities have simply ignored Homann's factual data.

Precession of ϕ' about the $\underline{\mathbf{k}}_o$ axis does not determine the spin obliquity ν in the body frame, as has been claimed by astronomers. It is instead nutation of γ' which alters ν for the Earth. Nutation is separate from ϕ' and should be compared to a spinning top which slowly nods as it precesses.

In Appendix 7 we have considered precession and nutation exercises for Mars, based on data obtained from space and landing craft which were sent there some years ago. The computations can only be estimated because we do not possess any reliable measurements for the nutation of the planet's spin axis over a period of at least one Martian year. The cited missions do provide a measurement of its precession, and we can formulate W/W_m for Mars. However, we disagree with the validity of estimates for the planet's principal body moment of inertia without some knowledge of the current values of γ' and γ''.

8 – THE PERIHELION ADVANCE OF MERCURY'S ORBIT

An anomalous component of the perihelion advance of Mercury's orbit was first noted in 1859 by Urbain Le Verrier (1811-1877). About 0.38 arc-seconds per year remained after 5.32 arc-seconds had been attributed to the gravity fields of other planets. The importance of the discrepancy is evidenced by diverse theories offered during that era, including modifications to the inverse square law and claims of the existence of an undiscovered planet. We shall address three proposed explanations for the anomalous component.

Perihelion Advance Based On The General Theory

In 1915, following several unsuccessful efforts, Albert Einstein proposed a modification of the gravitational force equation based on the general theory of relativity. His approach claims that the solar mass of $M \cong 1.991 \times 10^{30}$ kilograms warps space in its vicinity and requires a planet to travel further than it would in Euclidean (flat) space in order to return to its perihelion. In Euclidean space the square of the distance ds between two points such as $\underline{\mathbf{r}}_1 = x_1\,\underline{\mathbf{i}} + y_1\,\underline{\mathbf{j}} + z_1\,\underline{\mathbf{k}}$ and $\underline{\mathbf{r}}_2 = x_2\,\underline{\mathbf{i}} + y_2\,\underline{\mathbf{j}} + z_2\,\underline{\mathbf{k}}$ is given by $ds^2 = dx^2 + dy^2 + dz^2 = dr^2 + r^2\,(d\varphi^2 + \sin^2\varphi\, d\theta^2)$, where $dx^2 = (x_1 - x_2)^2$, etc. In the general theory a spherically symmetric distance element is modeled by equation 1, where $R_z = 2Mg_o/c^2$. Inputting the gravitational constant $g_o \cong 6.670 \times 10^{-20}$ kilometers cubed per second squared per kilogram and the speed of light $c \cong 2.998 \times 10^5$ kilometers per second, we find that $R_z \cong 2.956$ kilometers for the sun. † Einstein named R_z *the Schwarzschild radius* after his friend who supported the development of the general theory, but died during World War I.

$$ds^2 = dr^2/(1 - R_z/r) + r^2\,(d\varphi^2 + \sin^2\varphi\, d\theta^2), \qquad \text{for } R_z = 2Mg_o/c^2. \qquad (1)$$

The factor $1/(1 - R_z/r)$ provides a small term, expression 2, for inclusion in the force equation, where r is a planet's distance from the sun, μ_p is its

---

† For extremely dense stars one might suppose that the entire mass lies inside the Schwarzschild radius, so that the expressions for distance and time-keeping become indeterminate at $r = R_s$. This is the basis for much publicized "black holes".

mass, $\mu_p J_{orb}$ is its orbital angular momentum, and $\underline{\hat{\mathbf{u}}}_r$ is a unit vector in the direction of $\underline{\mathbf{r}}$. The force modification is less than one-millionth of the average solar attraction of 3.96×10^{-5} kilometers per second-squared for Mercury. †

$$\underline{\mathbf{F}}_p = -3\mu_p M g_o J_{orb}^2\, \underline{\hat{\mathbf{u}}}_r/(c^2 r^4), \qquad \text{Einstein force perturbation,}$$

$$F_p/\mu_p \cong -2.945 \times 10^{-12} \text{ kilometers per second-squared,} \tag{2}$$

$$\begin{aligned} d^2u/d\theta^2 + u &= 1 + 3u^2 M^2 g_o^2/(J_{orb}^2 c^2) \\ &= 1 + 3\,(1 + 2\varepsilon \cos\theta + \varepsilon^2 \cos^2\theta)\, M^2 g_o^2/(J_{orb}^2 c^2), \\ &\qquad \text{where } u = J_{orb}^2/(Mg_o r), \text{ and } \cos\varphi = -\sin\iota \cos\theta, \end{aligned} \tag{3}$$

$$u = 1 + \varepsilon \cos\theta + [3M^2 g_o^2/(J_{orb}^2 c^2)]\,[1 + \varepsilon\theta \sin\theta + \varepsilon^2 (3 - \cos 2\theta)/6)], \tag{4}$$

$$r = (J_{orb}^2/Mg_o)/\{1 + \varepsilon \cos[1 - 3M^2 g_o^2/(J_{orb}^2 c^2)]\,\theta + o(\varepsilon^2)\}. \tag{5}$$

The inclusion of $-3\mu_p M g_o J_{orb}^2/(c^2 r^4)$ in radial equation 2-4a of Appendix 2 results in equation 3 for $u = J_{orb}^2/(Mg_o r)$. The unperturbed solution is, as before, $u_o = 1 + \varepsilon \cos\theta$, where ε is the orbit eccentricity and θ is the angular position in a sun-centered frame whose polar axis is aligned with $\underline{\mathbf{J}}_{orb}$. Inputting $u \cong u_o$ to its right side, we obtain the second line of equation 3.

Since the term $\cos\theta$ is a general solution of equation 3, the form $\cos\theta$ multiplied by a constant will not provide the particular solution required by the perturbation. However, inputting expression 4 to $d^2u/d\theta^2 + u$ produces the term $6\varepsilon (\cos\theta)\, M^2 g_o^2/(J_{orb}^2 c^2)$ on the left side of equation 3. All terms multiplying $3M^2 g_o^2/(J_{orb}^2 c^2)$ in the solution are constant or oscillatory except $\theta \sin\theta$, which slowly increases with θ. Since $3M^2 g_o^2\, \theta/(J_{orb}^2 c^2)$ is so small in equation 4, we may substitute $\sin[3M^2 g_o^2 \theta/(J_{orb}^2 c^2)]$ for $3M^2 g_o^2 \theta/(J_{orb}^2 c^2)$, replace $\varepsilon \cos\theta$ by $\varepsilon \cos\theta \cos[3M^2 g_o^2 \theta/(J_{orb}^2 c^2)]$, and set $1 + 3M^2 g_o^2/(J_{orb}^2 c^2) \cong 1$.

Using the identity $\cos(\alpha + \beta) = \cos\alpha \cos\beta - \sin\alpha \sin\beta$, we obtain expression 5 as the physically significant solution for r, approximately representing an ellipse where θ must increase by the angle $2\pi\, [1 + 3M^2 g_o^2/(J_{orb}^2 c^2)]$ in

† See, Stephani, H., *General Relativity*, Cambridge University Press (1977, Eng. Tr. 1990), at pp. 103-108, for additional discussion of the perturbation.

order to return to its original perigee. Inputting $Mg_o = 1.328 \times 10^{11}$ kilometers-cubed per second-squared with $a_n(1 - \varepsilon^2) = J_{orb}{}^2/(Mg_o) = 0.5546 \times 10^8$ kilometers for Mercury's semi-major axis of $a_n = 0.5791 \times 10^8$ kilometers and eccentricity $\varepsilon = 0.2056$, we obtain $6\pi M^2 g_o{}^2/(J_{orb}{}^2 c^2) \cong 5.018 \times 10^{-7}$ radian per orbit $\cong 0.1035$ arc-seconds per orbit. Since Mercury's orbital period is only 0.241 that of the Earth, the rate is about 0.43 arc-seconds per Earth year. This result has been a major impetus for acceptance of the general theory.

Perihelion Rotation Due To $\underline{\mathbf{B}}_g$

The general theory perturbation looks a lot like the magnetic type of phenomenon we have modeled. Reviewing our above results, we found that $(Mg_o/a_o)^{1/2} = Mg_o/\sigma$ appears to be a universal gravitational constant of approximate value 68.25 kilometers per second, such that $\sigma \cong 1.946 \times 10^9$ kilometers-squared per second and $a_o \cong 0.285 \times 10^8$ kilometers for the solar gravitational field. Using a result of Maxwell's equations for electromagnetism as an analogy for gravity waves, we specified $\kappa_g \cong 9.324 \times 10^{-27}$ meters per kilogram, where $M\kappa_g/(4\pi)$ is identically Mg_o/c^2 and has a value of 1.478 kilometers, which is half of the Schwarzschild radius for the sun.

We have shown that whenever J_{orb} is modeled in a spin-stabilized frame inclined at an angle γ to $\underline{\mathbf{k}}_o$ in the $\underline{\mathbf{i}}_o\underline{\mathbf{k}}_o$ plane, the mass current ϑ due to the orbital motion encounters a torque proportional to $\underline{\mathbf{B}}_g = -\sigma(\cos\varphi)\,\underline{\mathbf{i}}/(2r^2)$, where $\underline{\mathbf{B}}_g = \nabla \times \underline{\mathbf{A}}_g$ for the gravitational vector potential $\underline{\mathbf{A}}_g$. But when the velocity vector lies in the $\underline{\mathbf{i}}_o\underline{\mathbf{j}}_o$ plane where $\cos\varphi = 0$, the torque due to the major component of the solar mass current vanishes. Nevertheless, a small torque due to a dipole moment for $\underline{\mathbf{B}}_g$ causes a slight, continuous rotation of an eccentric orbit about its vector, as we shall now demonstrate.

In Chapter 4 we allowed for a dipole term $\underline{\mathbf{p}}_1$ which leads to a flux of $\underline{\mathbf{B}}_{g1}$, expression 6, lying in the $\underline{\mathbf{i}}_o\underline{\mathbf{j}}_o$ plane. The phenomenon is comparable to a scalar potential V for which oppositely signed charges of q create the dipole potential $V_1 = \underline{\mathbf{p}} \bullet \underline{\hat{\mathbf{u}}}_r/(4\pi\epsilon_o r^2)$, where $\underline{\mathbf{p}} = q\,\Delta\underline{\mathbf{x}}$ and $\Delta\underline{\mathbf{x}}$ is the separation of the charges. Similarly, $\underline{\mathbf{B}}_{g1}$ creates a non central force of $\underline{\mathbf{F}}_{B1} = \mu_p \underline{\mathbf{v}} \times \underline{\mathbf{B}}_{g1}$, where $\underline{\mathbf{v}}$ is the planet's velocity, and the force produces a torque $\underline{\boldsymbol{\tau}} = \mu_p\,\underline{\mathbf{r}} \times \underline{\mathbf{F}}_{B1}$, expression 7, where $\underline{\mathbf{a}} \times (\underline{\mathbf{b}} \times \underline{\mathbf{c}}) \equiv (\underline{\mathbf{a}} \bullet \underline{\mathbf{c}})\,\underline{\mathbf{b}} - (\underline{\mathbf{a}} \bullet \underline{\mathbf{b}})\,\underline{\mathbf{c}}$ for any vectors $\underline{\mathbf{a}}$, $\underline{\mathbf{b}}$, and $\underline{\mathbf{c}}$.

$$\underline{\mathbf{B}}_{g1} = -\sigma\,(p_1 \sin\varphi)\,\hat{\underline{\mathbf{u}}}_\theta / r^3, \tag{6}$$

$$\underline{\tau}/\mu_p = \underline{\mathbf{r}} \times (\underline{\mathbf{v}} \times \underline{\mathbf{B}}_{g1}) \equiv (\underline{\mathbf{r}} \cdot \underline{\mathbf{B}}_{g1})\,\underline{\mathbf{v}} - (\underline{\mathbf{r}} \cdot \underline{\mathbf{v}})\,\underline{\mathbf{B}}_{g1} = (\underline{\mathbf{r}} \cdot \underline{\mathbf{v}})\,\sigma\,(p_1 \sin\varphi)\,\underline{\mathbf{u}}_\theta / r^3. \tag{7}$$

When we compute the torque $\underline{\tau}$ on the orbit, the term $\underline{\mathbf{r}} \cdot \underline{\mathbf{B}}_{g1}$ vanishes since $\underline{\mathbf{B}}_{g1}$ lies in the $\hat{\underline{\mathbf{u}}}_\theta$ direction and $\underline{\mathbf{r}} \cdot \hat{\underline{\mathbf{u}}}_\theta$ is zero. The second term also vanishes for circular orbits since $\underline{\mathbf{r}} \cdot \underline{\mathbf{v}}$ is zero. † Thus, the only non vanishing torque in our model that impacts orbital motion in the $\underline{\mathbf{i}}_o\underline{\mathbf{j}}_o$ plane is aligned with $\underline{\mathbf{u}}_\theta$, expression 7, and affects only eccentric orbits such as Mercury's.

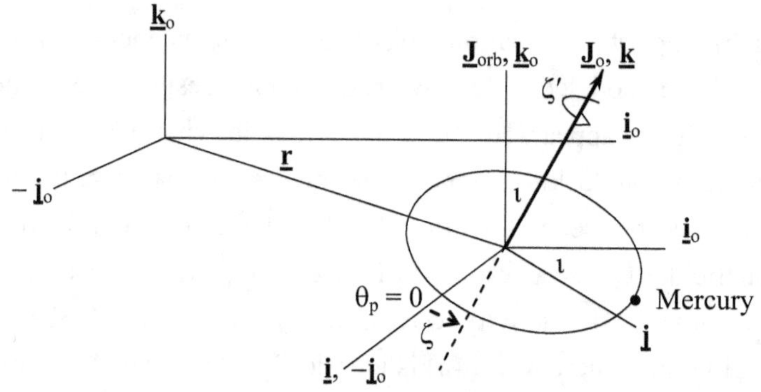

Figure 8.1. Small Orbit-Level Frame Rotation For Mercury

Figure 8.1 depicts Mercury's orbit as it appears in an inertial solar $\underline{\mathbf{i}}_o\underline{\mathbf{j}}_o\underline{\mathbf{k}}_o$ frame and also in an $\underline{\mathbf{ijk}}$ frame where $\underline{\mathbf{k}}$ is aligned with $\underline{\mathbf{J}}_o$. We have shown that Mercury's specific angular momentum is $\underline{\mathbf{J}}_{orb} = 2^{1/2}\,\sigma\,\underline{\mathbf{k}}_o \cong J_o \cos\iota\,\underline{\mathbf{k}}_o$, lying along the $\underline{\mathbf{k}}_o$ axis. The actual velocity vector $\underline{\mathbf{v}}$ lies in the $\underline{\mathbf{i}}_o\underline{\mathbf{j}}_o$ plane, even though the observed orbit vector along the $\underline{\mathbf{k}}$ axis is inclined to $\underline{\mathbf{k}}_o$ at $\iota \cong 18.4$ degrees due to the effect of $\underline{\mathbf{A}}_g$. The $\underline{\mathbf{j}}$ axis is perpendicular to $\underline{\mathbf{J}}_o$, and initially lies in the $\underline{\mathbf{i}}_o\underline{\mathbf{k}}_o$ plane, so that the orbit perihelion θ_p lies along the $\underline{\mathbf{i}}$ axis, *i.e.*, $\theta_p = 0$. †† We shall model frame rotation as a small value of ζ' about $\underline{\mathbf{k}}$, which produces an advance of the perihelion in the direction of the orbital motion

† See, Kibble, T.W.B., *Classical Mechanics*, Wiley & Sons, NY (1973), pp. 83-87, for an analysis of Larmor precession.

†† The initial value of θ_p is simply a reference point for the rotation of the orbit, and does not affect the end result.

by the angle ζ. In the **ijk** frame, r and $r' = dr/dt$ are given by expressions 8 and 9, so that $\underline{\mathbf{r}} \bullet \underline{\mathbf{v}} = r\,r' = J_{orb}\,\varepsilon\,(\sin\theta)/(1+\varepsilon\cos\theta)$ in expression 7.

$$r = a_n(1-\varepsilon^2)/(1+\varepsilon\cos\theta), \quad \text{where } a_n \text{ is the orbit's semi-major axis, and } \varepsilon \text{ is its eccentricity,} \tag{8}$$

$$r' = r\,\varepsilon\,\theta'(\sin\theta)/(1+\varepsilon\cos\theta), \quad \text{where } \theta' = J_{orb}/r^2, \tag{9}$$

$$\underline{\tau}/\mu_p = [J_{orb}\,\varepsilon/(1+\varepsilon\cos\theta)]\,\varpi\,(p_1\sin\varphi)\,(\sin^2\theta\,\underline{\mathbf{i}} - \sin\theta\cos\theta\,\underline{\mathbf{j}})/r^3$$
$$\cong J_{orb}\,\varepsilon\,\varpi\,(p_1\sin\varphi_{av})\,\underline{\mathbf{i}}/(2r_{av}^3), \quad \text{for } \sin\varphi_{av} \cong 0.975, \tag{10}$$

$$\zeta'\,\underline{\mathbf{k}}\times\underline{\mathbf{J}}_{orb} = \zeta'\,J_{orb}\,(\sin\iota)\,\underline{\mathbf{i}} \cong J_{orb}\,\varepsilon\,\varpi\,(p_1\sin\varphi_{av})\,\underline{\mathbf{i}}/(2r_{av}^3), \tag{11}$$

$$\zeta' \cong \varepsilon\,\varpi\,(p_1\sin\varphi_{av})/(2r_{av}^3\sin\iota)$$
$$\cong 6.179\,(p_1/r_{av}) \text{ radian per year.} \tag{12}$$

Substituting from expression 7, equation 10 specifies the torque on the orbit. Replacing $\sin^2\theta$ by $(1-\cos 2\theta)/2$ and averaging over the orbit, the **i** component averages to zero, but $\sin^2\theta$ averages to 1/2 and leads to the end form of equation 10. Since $\sin\varphi$ is never negative and varies between $\cos\iota = 0.95$ and 1, let us use the average of $\sin\varphi_{av} \cong 0.975$ with $r_{av} \cong 0.5791 \times 10^8$ kilometers. Equating $\underline{\tau}/\mu_p$ to $\zeta'\,\underline{\mathbf{k}}\times\underline{\mathbf{J}}_{orb}$, we obtain equation 11. Cancelling common J_{orb} factors leads to equation 12, and using 1 year $\cong 3.156 \times 10^8$ seconds to express ϖ, we obtain the end form when we input ε, ι, and ϖ/r_{av}^2.

If we estimate the "freely adjustable" value of p_1 at about 6.6 times the Schwarzschild radius, we obtain the same result that Einstein's theory provides. The magnetic-like dipole result has several advantages, including (1) the adjustability of p_1 to agree with observation data obtained at a future date for a non-zero quadrupole, and (2) the consistency of the dipole approach with the magnetic-like force's ability to explain major planetary orbit parameters, including the inclinations, spin states, and mean orbit radii. Moreover, the dipole approach predicts a negligible effect on Venus' orbit due to its almost zero eccentricity, and very little for the Earth and Mars, consistent with observations. In contrast, the Einstein theory is independent of orbit eccentricity and applies at levels not observed for Venus, Earth, and Mars.

Although the results afforded by these two theories are of the same order of magnitude, their natures are vastly different. Ours is the gravitational equivalent of *Larmor precession* in electromagnetic theory and has nothing to do with a postulated warping of space by the solar field. Observations of perihelion advances for spacecraft in orbits of varying eccentricities and inclinations about the sun should determine the correctness of the approaches.

The Quadrupole Solution

In Chapter 7 above we considered the oblateness of the Earth and modeled the effects of non central gravitational forces on its stabilized spin axis. The estimate of the Earth's oblateness is based on observations of artificial satellite orbits with perturbations modeled as being due to a second order geopotential $V_{02}(r_e)$, expression 13, where r_e is the distance from the satellite to the Earth's center, $J_2 = (I_s - I_a)/(\mu h_a^2) \cong 1082.6 \times 10^{-6}$, λ is latitude on the Earth, and $h_a \cong 6{,}378$ kilometers for the Earth's mean equatorial radius. †

$$V_{02}(r_e) = (\mu g_o/r_e)\,[1 - J_2(h_a/r_e)^2\,(3\sin^2\lambda - 1)/2]. \tag{13}$$

One effect of J_2 is to create precession of a satellite's orbit, causing its intersection with the Earth's equator to drift slowly westward on successive orbits. The precession decreases as the orbit inclination increases, becoming zero at 90 degrees inclination and reversing the drift direction at inclinations from 90 to 180 degrees. A second effect is to cause the orbit perigee to rotate in the direction of the satellite's motion at inclinations less than about 63.4 degrees, to cease the effect at this value, and to reverse the perigee rotation direction at higher inclinations. This behavior applies for a quadrupole in any central attracting mass, and we derive it below using spherical polar coordinates instead of the specialized coordinates that astronomers use.

Let us ignore the effects of $\underline{\mathbf{A}}_g$ in this exercise and consider the orbit geometry illustrated in Figure 8.2, where a quadrupole moment lies in the $\underline{\mathbf{ij}}$ plane, and the orbit vector $\underline{\mathbf{J}}_{orb}$ is inclined to that plane at an angle α which might not be ι. We are measuring the azimuth angle θ relative to the $\underline{\mathbf{i}}$ axis,

---

† See, Bate, et al., *Fundamentals of Astrodynamics*, *supra*, p. 422.

with the complementary elevation angle φ between **k** and the planet's position **r** being given by $\cos\varphi = -\sin\alpha\,\cos\theta$. Since we are concerned only with averages, our results are independent of the perigee location.

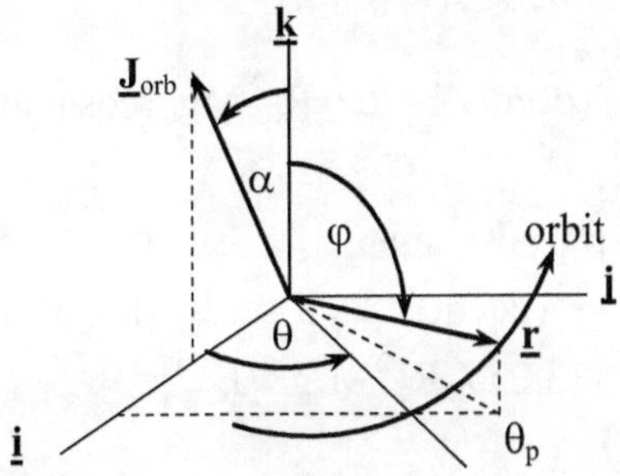

Figure 8.2. Orbit Inclined To A Quadrupole Plane †

The scalar potential V_{04} due to the quadrupole moment is provided by Laplace's solution, expression 14. Observations determine C_4, and following convention, we have expressed C_4 as $Mg_o J_2\, R_S^{\,2}$, where M is the sun's mass, R_S is its equatorial radius, and J_2 is a dimensionless constant. Expression 15 is a Lagrangian which includes V_{04}.

$$V_{04} = \mu\, C_4\,(3\cos^2\varphi - 1)/(2r^3) = \mu J_2\, R_S^{\,2}\, Mg_o\,(3\cos^2\varphi - 1)/(2r^3), \qquad (14)$$

$$L = \mu\,[(dr/dt)^2 + r^2\,(d\varphi/dt)^2 + r^2\,(d\theta/dt)^2\sin^2\varphi]/2 + \mu Mg_o/r - V_{04}. \qquad (15)$$

Using the Lagrangian methodology, expression 16 specifies the planet's angular momentum along the **k** axis, which is not affected by the quadrupole. Equation 17 is the φ equation of motion. Defining the dimensionless variables $u = J_{orb}^{\,2}/(Mg_o r)$ and $q = r^2\,(d\varphi/dt)/J_{orb}$, with the small dimensionless parameter $\beta = (3/2)\,J_2\,R_S^{\,2}\,(M^2 g_o^{\,2}/J_{orb}^{\,4})$, we obtain equation 17a. When J_2 is zero, the solutions are given by $u_o = 1 - \varepsilon\cos\varphi/\sin\alpha$ and $q_o^{\,2} = 1 - \cos^2\alpha/\sin^2\varphi$. Retaining only the lowest order term for β, *i.e.*, setting $u = u_o$ in equation

---

† Although we are making theory comparisons based on the assumption that α is ι = 18.4 degrees for Mercury's orbit, the mass quadrupole could lie in a plane other than the $\underline{i}_o\underline{i}_o$ plane of our theory. The **k** axis does not appear to be the sun's plasmatic north pole, which is inclined to the ecliptic at about 7 degrees.

17a, we obtain equation 17b. Upon integrating, expression 17c provides the solution for $q^2 = r^4 (d\varphi/dt)^2/J_{orb}^2$, where b_o is a constant.

$$r^2 (\sin^2\varphi)\, d\theta/dt = J_{orb} \cos\alpha, \qquad (16)$$

$$d/d\varphi\, [r^4 (d\varphi/dt)^2 + r^4 (\sin^2\varphi)(d\theta/dt)^2] = 6J_2 R_S^2 Mg_o (\cos\varphi \sin\varphi)/r, \qquad (17)$$

$$d/d\varphi\, (q^2 + \cos^2\alpha/\sin^2\varphi) = 4\beta\, u \cos\varphi \sin\varphi,$$
$$\text{where}\quad q = r^2 (d\varphi/dt)/J_{orb},\ u = J_{orb}^2/(Mg_o r), \text{ and}$$
$$\beta = (3/2) J_2 R_S^2 M^2 g_o^2/J_{orb}^4, \qquad (17a)$$

$$q\, dq/d\varphi = \cos^2\alpha \cos\varphi/\sin^3\varphi + 2\beta\, (1 - \varepsilon \cos\varphi/\sin\alpha) \cos\varphi \sin\varphi,$$
$$\text{where}\quad u_o = 1 - \varepsilon \cos\varphi/\sin\alpha, \qquad (17b)$$

$$q^2 = r^4 (d\varphi/dt)^2/J_{orb}^2 = 1 - \cos^2\alpha/\sin^2\varphi + 2\beta\, g(\varphi)],$$
$$\text{where } g(\varphi) = [b_o - \cos^2\varphi + 2\varepsilon (\cos^3\varphi)/(3\sin\alpha)]. \qquad (17c)$$

Equation 18 is the r-equation of motion. Expressing $dr/d\varphi\ d\varphi/dt$ as dr/dt, we have $dr/dt = dr/d\varphi\ J_{orb}\, q/r^2 = Mg_o\, q\, (-du/d\varphi)/(J_{orb}\, u^2)$. With effort we find that $d^2r/dt^2 = -\,(M^3 g_o^3/J_{orb}^4)\, u^2\, (q^2\, d^2u/d\varphi^2 + q\, du/d\varphi\, dq/d\varphi)$. Inputting this result to equation 18 and dividing by $(M^3 g_o^3/J_{orb}^4)\, u^2$, which cannot be zero, we obtain equation 18a, where $J_{orb}^2/Mg_o = a_n (1- \varepsilon^2)$, for the orbit's semi-major axis a_n and eccentricity ε. When we input the solution form $u = u_o + \beta\, u_1$ to equation 18a and retain only first-order terms having β as a common multiplier, equation 18b results. Terms not multiplied by β cancel since u_o is the solution when $\beta = 0$, and those containing u and q multiplied by β have been approximated as βu_o and βq_o. We have also transposed the multipliers of u_o to the right side of the equation.

$$d^2r/dt^2 - [r^2 (d\varphi/dt)^2 + r^2 (d\theta/dt)^2 \sin^2\varphi]/r^3 + Mg_o/r^2$$
$$= 3J_2 R_S^2 Mg_o (3\cos^2\varphi - 1)/(2r^4), \quad \text{where } r = J_{orb}^2/(Mg_o u), \qquad (18)$$

$$q^2\, d^2u/d\varphi^2 + q\, du/d\varphi\, dq/d\varphi + [1 + 2\beta\, g(\varphi)]\, u = 1 - \beta\, (3\cos^2\varphi - 1)\, u_o^2,$$
$$\text{where } u = u_o + \beta\, u_1, \text{ and } u_o = 1 - \varepsilon \cos\varphi/\sin\alpha, \qquad (18a)$$

$$d^2u_1/d\varphi^2\,(1 - \cos^2\alpha/\sin^2\varphi) + du_1/d\varphi\,(\cos^2\alpha\,\cos\varphi/\sin^3\varphi) + u_1 =$$
$$(1 - 2b_o) - (4\varepsilon/\sin\alpha)\,[\cos\varphi - 5(\cos^3\varphi)/3]$$
$$+ (3\varepsilon^2/\sin^2\alpha - 1)\cos^2\varphi - 5\varepsilon^2\sin^2\alpha\cos^4\varphi. \qquad (18b)$$

Substituting $\cos\varphi = -\sin\alpha\cos\theta$, we obtain the first form of equation 19. Using multiple angle identities, especially $\cos^3\theta = (3\cos\theta + \cos3\theta)/4$, we obtain the second form of equation 19, where we have set the arbitrary constant $2b_o = 1 + 3\varepsilon^2/2 - (\sin^2\alpha)/2 - 15\,\varepsilon^2(\sin^2\alpha)/8$ to facilitate the computation. Inputting $u_{1\varepsilon} = (2 - 2.5\sin^2\alpha)\,\varepsilon\,\theta\sin\theta$ to $d^2u_1/d\theta^2 + u_1$ as a component of u_1, we obtain a term of $\varepsilon(4 - 5\sin^2\alpha)\cos\theta$ on the right side of the equation.

$$d^2u_1/d\theta^2 + u_1 = (1 - 2b_o) + 4\varepsilon\,[\cos\theta - 5\sin^2\alpha\,(\cos^3\theta)/3]$$
$$+ (3\varepsilon^2 - \sin^2\alpha)\cos^2\theta - 5\varepsilon^2\sin^2\alpha\cos^4\theta$$
$$= 1 + 3\varepsilon^2/2 - (\sin^2\alpha)/2 - 15\,\varepsilon^2(\sin^2\alpha)/8 - 2b_o$$
$$+ \varepsilon(4 - 5\sin^2\alpha)\cos\theta - 5\varepsilon\,(\sin^2\alpha\cos3\theta)/3$$
$$+ (3\varepsilon^2 - \sin^2\alpha - 5\varepsilon^2\sin^2\alpha)(\cos2\theta)/2$$
$$- 5\varepsilon^2(\sin^2\alpha\cos4\theta)/8. \qquad (19)$$

Since the remaining terms are oscillations over the orbit cycle, expression 20 is the physically significant solution. Expressing $\beta\,(2 - 2.5\sin^2\alpha)\,\theta$ as $\sin[\beta\,(2 - 2.5\sin^2\alpha)\,\theta]$ and $\cos[\beta\,(2 - 2.5\sin^2\alpha)\,\theta]$ as 1 for small values of β and using the cosine sum identity, the solution takes the end form of expression 20. The complete solution for r is then given by expression 21.

$$u = u_o + \beta\,u_1 = 1 + \varepsilon\cos\theta + \varepsilon\,\beta\,(2 - 2.5\sin^2\alpha)\,\theta\sin\theta,$$
$$= 1 + \varepsilon\cos[1 - \beta\,(2 - 2.5\sin^2\alpha)]\,\theta, \qquad (20)$$

$$r = (J_{orb}^2/Mg_o)/\{1 + \varepsilon\cos[1 - 3\,(J_2R_S^2M^2g_o^2/J_{orb}^4)(1 - 1.25\sin^2\alpha)]\,\theta\}. \quad (21)$$

Expression 21 is a classical result, showing that the perihelion advance varies according to $1 - 1.25\sin^2\alpha$ for the inclination angle α, which specifies the critical value $\alpha = \text{arc-}\sin(1/1.25)^{1/2} \cong 63.4$ degrees. The expression for r models an ellipse which is slowly rotating in a counterclockwise direction, where θ must increase by an angle $2\pi\,[1 + 3\,(1 - 1.25\sin^2\alpha)\,J_2R_S^2M^2g_o^2/J_{orb}^4]$ during any revolution in order to return to the original perigee location.

When $\sin^2\alpha = 1/1.25$, *i.e.*, $\alpha \cong 63.44$ degrees, the rotation ceases, and when $\sin^2\alpha > 1/1.25$, the multiplier of θ becomes negative and the perigee rotates in a clockwise direction, as is observed for artificial Earth satellites. Russia has used this inclination since the 1960s to stabilize Molniya communication satellites, which are placed in 12-hour Earth orbits with eccentricities of about 0.7 to maximize the time spent above the northern hemisphere.

Although visual observations of the sun's outer surface indicate a spherical shape, an overview of the solar system as a whole shows a discoid with the sun at its center. In addition, spectral analyses of solar radiation indicate the existence of heavy elements within the sun in quantities comparable to those in the Earth. We would therefore expect a slightly higher mass density in an equatorial solar plane due to the sun's rotation, and the failure to observe an equatorial bulge on its volatile surface does not disprove the existence of a quadrupole moment within the solar mass.

It would not be unreasonable to adopt a solar J_2 value of roughly one-sixth of the Earth's value due to a mass aggregation in its equator. If we set the solar value of J_2 to 0.193×10^{-3} for a body radius of 695,700 kilometers and take J_{orb}^2 to be approximately 7.17×10^{18} kilometers to the fourth power per second-squared for Mercury's orbit, the advance predicted by expression 21 becomes 4.98×10^{-7} radian per orbit, or 0.43 arc-seconds per year. †

Theory Comparisons

Using a common base value for Mercury's advance and normalizing to the Earth's orbit period with factors of 0.241, 0.615, and 1.881 for the periods of the other terrestrial planets, Table 6 compares the predictions for the three proposed theories. Since r is proportional to J_{orb}^2, the general relativity values in equation 5 decrease as 1/r, whereas the quadrupole effect due to $1/J_{orb}^4$ falls off as $1/r^2$. Neither of these varies as an explicit function

† The value of J_2 proposed by general theorists for the sun in available literature is less than one percent of the Earth's value. See, *e.g.*, Biswas, A., and Mani, K., *Relativistic Perihelion Precession of Orbits of Venus and the Earth*, Dept. of Physics, Godopy Center, Calcutta, India, providing relativistic predictions based on an alternate relativistic gravitation model that relies on JPL's ephemeris DE405.

of orbit eccentricity. Only the magnetic-like dipole effect so varies, consistent with the failure to observe significant perihelion advances for planets other than Mercury.

Table 6. Perihelion Advance: Arc-seconds (″) Per Earth Century
1 Astronomical Unit (a.u.) ≅ 1.496×10^8 kilometers

Planet	Orbit axis: a_n (a.u.)	Orbit eccen: ε	Orbit incl: ι (deg.)	Gen Rel advance (arc-sec)	Mag Dipl advance (arc-sec)	Quadpl advance (arc-sec)
Mercury	0.3871	0.2056	18.43°	42.9″	42.9″	42.9″
Venus	0.7233	0.0068	14.04°	8.62″	0.28″	4.71″
Earth	1.0000	0.0168	11.31°	3.84″	0.33″	1.56″
Mars	1.5237	0.0933	9.46°	2.42″	0.61″	0.37″

For terrestrial planet orbits other than Mercury's, the general relativity predictions are by far the greatest of the three approaches. However, verified observational data are not available for the other terrestrial planets even though space age technology should have enabled accurate determinations by this time, especially for Venus and the Earth. Where are these observations? After all, Le Verrier found 38 arc-seconds of anomalous precession per century for Mercury from limited telescopic data in 1859. Telling tests would be observations as a function of orbit inclination for stable spacecraft orbits about the sun. † The most likely finding will be that the anomalous component is some mixture of the approaches; however, it is possible that any one of them may not apply to the solar field. On the other hand, the verification or rejection of any of the three approaches would not rule out the viability of our theory based on the principal component of $\underline{\mathbf{B}}_g$.

---

† The bending of stellar light rays as they pass near the sun during a solar eclipse provides a companion test for the general theory's applicability to the solar field. The two tests are not, of course, independent. Ray bending observations since 1919 have been shrouded in controversy, and some have indicated that the general theory does not pass this test. The physics community has generally discarded the failing tests as being experimentally flawed, and there seems to have been no consideration of the possibility that deflections under different viewing geometries might be angular dependent, which option the general theory does not allow.

9 – ANOMALOUS TRAJECTORIES OF PIONEER SPACECRAFT

Anomalies in the trajectories of Pioneers 10 and 11, observed from 1980 to 2002 following flybys of Jupiter and Saturn in the 1970s, are due at least in part to the existence of the vector potential $\underline{\mathbf{A}}_g$ in the solar field.

The Pioneer Data

Pioneer 10 left the Earth in March 1972 on a survey mission to Jupiter, arriving there in December 1973. After completing that task, the spacecraft received a boost from Jupiter's orbital motion and continued its outward journey, exiting the solar system in 2003 at a speed of about 10 kilometers per second. Pioneer 11 was launched in April 1973 and later obtained a boost from Jupiter which took it out of the ecliptic plane and inside Jupiter's orbit on a survey mission to Saturn. After moving beyond Saturn's orbit both Pioneers coasted for years without thrusting. During these periods the spacecraft experienced persistent, statistically significant decelerations in the direction of the sun with a reported mean, or estimated "canonical" value, of $r_c'' = 8.74\ (\pm 1.33) \times 10^{-10}$ meters per second squared. The decelerations were confirmed both by Doppler shifts in the signals and by timing signal returns. The trajectories deviated from Newtonian paths as much as 400 kilometers per year. Below is a brief narrative provided by JPL analysts. [†]

> "For Pioneer 10, an approximately constant anomalous acceleration seems to exist in the data as close in as 27 AU from the Sun. For Pioneer 11, beginning just after Jupiter flyby, the early navigational data show a small value for the anomaly during the Jupiter-Saturn cruise phase in the interior of the solar system. But right at Saturn encounter, when the craft passed into a hyperbolic escape orbit, there was an apparent fast increase in the anomaly, where after it settled into the canonical value. The trajectories of Pioneer 10 and 11 were profoundly different. After its encounter with Jupiter, Pioneer 10 continued on a hyperbolic escape trajectory, leaving the solar system while remaining close to the

---

† See, Turyshev, S. and Toth, V. (2010), "The Pioneer Anomaly", *Living Reviews*, Relativity, 13, 4; http://www.livingreviews.org/lrr-2010-4.

plane of the ecliptic. Pioneer 11, in contrast, proceeded from Jupiter to Saturn along a trajectory that took it closer to the Sun, while outside the ecliptic plane. After its encounter with Saturn, Pioneer 11 also proceeded along a hyperbolic escape trajectory, but once again it was flying outside the plane of the ecliptic. In the end, the two spacecraft were flying out of the solar system in approximately opposite directions."

The data show that the Pioneer anomaly is not properly characterized by a constant value of r_c'' at some heliocentric distance r, but is trajectory dependent. Both Pioneer spacecraft possessed sufficient energy to escape the solar system after departing Jupiter at r ≅ 5.20 AU (astronomical units), but solar radiation pressure and Jupiter's gravitational field prevented the trajectories from being purely hyperbolic and masked variations in the anomalies. The anomaly after Pioneer 11's encounter with Saturn showed a gradual rise from zero with large uncertainties. [†] Subsequent data analyses have shown the anomaly to be slowly decreasing with distance r, [††] and later reports have changed some values of r_c'' as a function of r.

Various sources have been proposed for the anomalous decelerations, including relativistic effects, electrical charging, dust, and interactions with onboard systems. The possibility of *new physics* was raised, but data analysts did not report any attempt to model a velocity-dependent gravitational force. All of the proposed sources were eventually rejected, except for a model of asymmetrical heat radiation by the spacecraft. The proposed heat sources were radioisotope thermoelectric generators (RTGs), [†††] whose radiation reflected from the back of the communications antenna pointed toward the Earth. The cited 2012 paper compares the results of independent thermal analyses with the Pioneer 10 data and concludes that heat radiation alone is

---

[†] See, Turyshev and Toth (2010), "The Pioneer Anomaly", *Living Reviews*, *supra*.

[††] See, Turyshev, Toth, Ellis, and Markwardt (2011), "Support for temporally varying behavior of the Pioneer anomaly from the extended Pioneer 10 and 11 Doppler data sets", *Physical Review Letters* **107** (8).

[†††] See, Turyshev, Toth, Kinsella, Lee, Lok, and Ellis (2012), "Support for the Thermal Origin of the Pioneer Anomaly", *Physical Review Letters* **108** (24).

responsible for the anomalies. However, the heat model cannot satisfactorily explain the abrupt onset and gradual rise of the anomalies, or the non canonical values of later launched and differently designed spacecraft.

Like Pioneers, some newer spacecraft such as Galileo and Ulysses rely on spin stabilization subsystems. These craft exhibited similar anomalies, but firm conclusions were not drawn from their data for various reasons, including proximity to the Sun. The Cassini spacecraft employs both reaction wheels and thrusters for attitude control, and relied on reaction wheels alone during its cruise phase, enabling precise acceleration measurements. Its nominal (average) unmodeled deceleration was $r_{nom}'' = 26.7\ (\pm 1.1) \times 10^{-10}$ meters per second squared, roughly three times as large as the mean Pioneer anomaly. However, the radiation of kilowatts of heat from RTGs mounted on the spacecraft body prevented resolution of the issue. Rosetta also exhibited anomalies during its Earth flybys, and the absence of RTGs ruled out thermal recoil pressure as a possible cause of the anomalies. †

Trajectories For The Pioneers

Relying on Doppler data collected by very large antennae of the Deep Space Network at sites in California, Australia, South Africa, and Spain, JPL analysts used a numerical ephemeris generation program to compute accurate trajectories for the Pioneers. The program involves thousands of lines of code and uses filtering techniques to smooth the data and reject outliers. The reduction of observational data over an interval of time provides solutions for the spacecraft positions and velocities in the form of state vectors based on statistical regression techniques. The modeling includes effects of planetary perturbations, interplanetary media, radiation pressure, general relativity, and bias and drift in the Doppler data. It also corrects for the

† See, Rievers and Laemmerzahl, "High precision thermal modeling of complex systems with application to the flyby and Pioneer anomaly" (2001), *Annalen der Physik* **523** (6): 439. See also, http://www.space.com/scienceastronomy/080229-spacecraft-anomaly.html, stating that velocities during Earth flyby for Galileo, Messenger, NEAR, Cassini, and Rosetta were measured within 0.1 millimeter per second by bouncing signals off the spacecraft.

Earth's body motion, including precession, nutation, sidereal rotation, polar motion, tidal effects, and tectonic plate drifts. A further feature is Kalman filtering to reduce random noise such as diurnal effects.

This level of detail was necessary to extract the anomalous deceleration component from the data, but we need not concentrate on them to compute first-order effects of $\underline{\mathbf{B}}_g$. The solar system exit velocity provides the total energy E, expression 1, which we can use to calculate the magnitude of the velocity v at a given distance r from the sun once the spacecraft was free of planetary influences. We have estimated the trajectory's specific angular momentum J by a reasonable fit of the observed values of r_c'', and its periapsis r_p, based on the estimated onset of the anomaly. The values of E and J^2 specify the hyperbola parameters for an in-plane angle η according to equations 2 and 3. They are similar to elliptical orbits, except that the sum of potential and kinetic energy is positive and the eccentricity ε is greater than 1.

$$E/\mu = v^2/2 - Mg_o/r, \qquad (1)$$

$$2EJ^2/(\mu M^2 g_o^2) = \varepsilon^2 - 1, \qquad (2)$$

$$r = (J^2/Mg_o)/(1 + \varepsilon\cos\eta), \qquad \text{for } \eta = \theta - \theta_p, \text{ and } \cos\varphi = -\sin\iota\cos\theta. \qquad (3)$$

Equation 3 provides the relationship between r and $\eta = \theta - \theta_p$ during free-space travel. The plane of the trajectory is inclined at the angle ι to the solar $\underline{\mathbf{k}}_o$ axis, and the solar azimuth reference angle θ is zero at the trajectory's lowest point in the solar frame. The periapsis $r_p = a_p(\varepsilon - 1)$ occurs at an azimuth angle θ_p. We have modeled only the free-space portion of the trajectory provided by expression 3 and used it to estimate the values of $r^2\varphi'$, which are needed for our computations.

Following the Jupiter encounters the spacecraft were essentially free of that planet's gravitational field by the time they had reached Saturn's orbit radius at $r \cong 9.54$ AU. The trajectory for Pioneer 10 was then hyperbolic, and Pioneer 11's trajectory became hyperbolic after it had escaped Saturn's field. Upon exiting the solar system, Pioneer 10 is expected to arrive at the star Aldebaran some 68 light years away in about 2 million years. Its total specific energy at this exit speed is $E/\mu = v_{ex}^2/2 \cong 50.4$ kilometers squared per second squared. † Using $Mg_o \cong 13.28 \times 10^{10}$ kilometers cubed per second squared in the classical relation $E/\mu = Mg_o/(2a_p)$, we obtain $a_p \cong 8.81$ AU.

Based on trial and error fitting, we propose a free-space trajectory for Pioneer 10 with a J value of 21.68×10^9 kilometers squared per second. The value of ε from equation 2 is 1.92, and equation 4 provides r as a function of η. The periapsis for this J value occurs at $r_p = a_p(\varepsilon - 1) \cong 8.10$ AU, where the trajectory velocity would be about 17.9 kilometers per second.

$$r = 23.66\,\text{AU}/(1 + 1.92\cos\eta), \quad \text{estimate for Pioneer 10.} \tag{4}$$

Perturbations Due To $\underline{\mathbf{B}}_g$

In our analysis we included in the solar field a magnetic-like force $\underline{\mathbf{F}}_B = \mu\,\underline{\mathbf{v}} \times \underline{\mathbf{B}}_g$, expression 5, where $\underline{\mathbf{v}}$ is the object's velocity and μ is its mass. Equations 6a, b, and c are the r, φ, and θ equations of motion for the trajectory, where $J^2 = r^4\varphi'^2 + r^4\theta'^2\sin^2\varphi - \varpi^2/4 = J_o^2 - \varpi^2/4$.

$$\underline{\mathbf{F}}_B = \mu\,\underline{\mathbf{v}} \times \underline{\mathbf{B}}_g = -\mu\varpi(\cos\varphi)(-r\varphi'\,\underline{\mathbf{u}}_r + r'\,\underline{\mathbf{u}}_\varphi)/(2r^2), \tag{5}$$

$$d^2r/dt^2 - J^2/r^3 + Mg_o/r^2 = \varpi r^2\varphi'(\cos\varphi)/(2r^3), \tag{6a}$$

$$[1/(2r^3\varphi')]\,[d/dt\,(r^4\varphi'^2 + r^4\theta'^2\sin^2\varphi)] = -\varpi r'(\cos\varphi)/(2r^2), \tag{6b}$$

$$d/dt\,(r^2\theta'\sin^2\varphi) = 0, \tag{6c}$$

$$r^4\varphi'^2 + r^4\theta'^2\sin^2\varphi = J_o^2 - \varpi\int r'\,r\cos\varphi\,d\varphi, \text{ where } J_o^2 = J^2 + \varpi^2/4, \tag{7b}$$

$$r^2\theta'\sin^2\varphi = J \cong J_o\cos\iota, \quad \text{where } \iota = \arctan[\varpi/(2J_o)] \cong 11.31 \text{ degrees.} \tag{7c}$$

Equation 6c is unaffected by $\underline{\mathbf{F}}_B$, but the r and φ equations include perturbations in the $\underline{\mathbf{u}}_\varphi$ and $\underline{\mathbf{u}}_r$ directions for small terms involving the constant $\varpi \cong 1.946 \times 10^9$ kilometers squared per second. Expressions 7b and c are the solutions for the φ and θ equations of motion. The trajectory plane

† See, http://www.solarviews.com/eng/pn10-11.htm, citing NASA data.

remains inclined to the $\underline{\mathbf{k}}_o$ axis of the solar field at approximately the angle ι.

The $\underline{\mathbf{u}}_\varphi$ perturbation produced by $\underline{\mathbf{B}}_g$ in equation 6b results in a small gradual increase in $r^4\varphi'^2 + r^4\theta'^2 \sin^2\varphi$ and a departure of the spacecraft position from its otherwise projected location. But we will ignore this contribution and treat the Doppler data as reflecting only the explicit dependence of equation 6a on $\underline{\mathbf{B}}_g$. Substituting $r^2\varphi' = J_o\,(1 - \cos^2\iota/\sin^2\varphi)^{1/2}$ and $\theta = \eta + \theta_p$ into 6a, where $\cos\varphi = -\sin\iota\,\cos\theta$, we obtain equation 8. Inputting σ and the average of $\sin\varphi_{av} = 0.99$ with $J_o \cong 21.70 \times 10^9$ kilometers squared per second, the anomaly r_{nom}'' takes its end form in kilometers per second squared at r kilometers. At r = 27 AU, the magnitude of r_c'' should be of the order 10^{-12} kilometers per second squared when $\sin 2(\eta + \theta_p) \cong 0.1$ or so.

$$\begin{aligned} r'' - J^2/r^3 + Mg_o/r^2 &= -\sigma J_o (1 - \cos^2\iota/\sin^2\varphi)^{1/2}\,[\sin\iota\,\cos(\eta + \theta_p)]/(2r^3) \\ &= -\sigma J_o \sin^2\iota\,[\sin 2(\eta + \theta_p)]/(4r^3 \sin\varphi) \\ &\cong -4.06 \times 10^{17}\,\mathrm{km}^4/\mathrm{sec}^2\,[\sin 2(\eta + \theta_p)]/r^3 = r_{nom}''. \qquad (8) \end{aligned}$$

The ratio of σJ_o to r^3 determines the magnitude of the anomaly r_c'' following its onset, which occurs at a value $r = r_a$ where $\eta + \theta_p = 0$. If, at a distance $r = r_b$, the $\sin 2(\eta + \theta_p)$ increases from zero at r_a in approximate proportion to $r_b{}^3/r_a{}^3$, the effect will appear to be nearly constant. This condition applies † for Pioneer 10, where the anomaly reported in 2002 was provided for 27 AU ≤ r ≤ 47 AU (from 1981 to 1989). A reasonable data fit occurs at $\theta_p \cong -89.8$ degrees for Pioneer 10, with the onset occurring at $r_a \cong 23.5$ AU and slowly rising to $r_c'' \cong -8.4 \times 10^{-10}$ meters per second at r = 27 AU.

Table 7a shows our theory results for r_c'' at the indicated solar separations and compares them with our best estimates for the data depicted by the graph in Figure 5.1 of the referenced 2010 report. Our findings should be acceptable in view of caveats in the reports. The r_c'' values provided by the 2002 report were based on a least squares data fit designed to provide a constant anomaly over an entire interval. Such an assumption obscures the

---

† See, Turyshev and Toth (2010), "The Pioneer Anomaly", *Living Reviews*, *supra*, at 83, citing Anderson, Laing, Lau, Liu, Nieto, and Turyshev, "Study of the anomalous acceleration of Pioneer 10 and 11", *Phys. Rev. D*, 65(8), 082004, (2002).

actual structure within the interval and is an average result. Another issue occurred when analysts estimated solar radiation acceleration at 18.9×10^{-10} meters per second squared for r = 10 AU and used a $1/r^2$ formula to average its effect over the interval 40 AU $\leq r \leq$ 70.5 AU, even though it varies by a factor of three in this interval. Moreover, the reported values "near the sun" are admittedly not actual measurements.

Table 7a. Pioneer 10 Anomaly For Periapsis At $\theta_p = -89.8$ Degrees

Solar Distance: r in AU	Reference Angle: η in degrees	r_c'' Theory: 10^{-10} m/sec^2	Original r_c'' Data: 10^{-10} m/sec^2
27 (1981)	93.7	–7.9	–8.8 ± 0.5
29	95.9	–10.2	–9.0 ± 0.2
31.5	97.4	–9.9	–8.5 ± 0.2
35.7	100.1	–9.2	–7.8 ± 0.2
40.5	102.5	–7.7	–7.7 ± 0.3
45.7 (1988)	104.5	–6.2	–7.0 ± 0.1

Turning to Pioneer 11, we are told that the spacecraft sent its last coherent Doppler data at r = 31.7 AU and that "As of the end of 1995 [Pioneer 11] was located at 44.7 AU from the Sun … heading outward at 2.5 AU/year." † Computing the total specific energy at this last location, we again find that $E/\mu = v_{ex}^2/2 \cong 50.4$ kilometers squared per second squared. The Jupiter flyby brought the trajectory inside Jupiter's orbit on its way to Saturn, and the flyby of Saturn again altered the parameters. After further trial and error, we have located the hypothetical periapsis for Pioneer 11 at $r \cong 1.32$ AU, where v_p would be 38.0 kilometers per second under the influence of the solar field alone and $J \cong 7.51 \times 10^9$ kilometers squared per second. We will use equation 9 for a hypothetical trajectory of Pioneer 11.

$$r = 2.84\text{ AU}/(1 + 1.15\cos\eta), \text{ trajectory estimate for Pioneer 11.} \qquad (9)$$

† See, http://nssdc.gsfc.nasa.gov/nmc/spacecraftDisplay.do?id=1973-019A. See also, Turyshev and Toth (2010), "The Pioneer Anomaly", *Living Reviews, supra.*

Table 7b. Pioneer 11 Anomaly For Periapsis At $\theta_p = -133.5$ Degrees

Solar Distance: r in AU	Reference Angle: η in degrees	r_c'' Theory: 10^{-10} m/sec^2	Original r_c'' Data: 10^{-10} m/sec^2 †
12.5 (1983)	132.3	n/a	−6.2 ± 1.8
13.6	133.5	0.0	−8.0 ± 2.1
17.0 (1985)	136.5	−9.0	−8.0 ± 0.9
18.2	137.3	−9.2	−9.0 ± 0.5
22.5	139.5	−7.7	−8.0 ± 1.0
23.0	139.7	−7.4	−9.0 ± 0.3
26.0	140.8	−6.0	−8.5 ± 0.1
29.0 (1990)	141.7	−4.9	−8.3 ± 0.2

Table 7b summarizes our results for Pioneer 11, where we have set θ_p to −133.5 degrees with $r_c'' = 0$ at $r \cong 13.6$ AU. The data were estimated from Figure 5.1 of the 2010 report, where some of the values indicate a lot of noise and large error bounds. However, the gradually rising form of r_c'' evidenced by the observational data supports our basic model. We do not expect our model to apply for values of r less than 17 AU since the lack of actual data as the spacecraft departs Saturn and is traveling faster than our model predicts causes us to overestimate r_a for the onset of the anomaly.

An article provided in 2014 combines subsequently retrieved data for Pioneer 10 with that for Pioneer 11 and specifies r_c'' at steps of 5 to 10 AU as r ranges from 20 AU to 70 AU. †† The article does not report any Pioneer 11 data prior to r = 20 AU and assumes that r_c'' remains nearly constant in the range $20 \text{ AU} \leq r \leq 30$ AU. The author estimates r_c'' in units of 10^{-10} meters per second squared at −7.0 for r = 50 AU and at −5.8 for r = 79.9 AU, respectively, and models r_c'' to be decreasing roughly as 1/r. However, there are no Doppler data for Pioneer 11 past r = 31.7 AU. Our value of r_c''

† We do not know the extent of Pioneer 11's being out of the ecliptic plane at the time its anomalous decelerations were observed, but have estimated its trajectory to lie roughly in the ecliptic based on the graph provided by Turyshev and Toth (2010), "The Pioneer Anomaly", *Living Reviews*, *supra*, at page 14, Figure 2.1.

†† See, https://spinor.info/weblog/?page_id=95, Toth, June 2014.

becomes -1.1×10^{-10} meters per second at r = 79.9 AU, and our model does not allow merger of the hyperbolic trajectories with differing parameters.

Attempts of the Pioneer data analysts to keep r_c'' constant within an interval, reported data uncertainties, and conflicts among the reports cast doubts on the viability of the heat radiation model. But the major shortcoming of the RTG model is its inapplicability for trajectories other than Pioneers. In contrast, the magnetic-like force $\underline{\mathbf{B}}_g$ affords variable results which can provide the observed data for other spacecraft when different trajectory parameters are input. Our model also offers a rational explanation for the anomaly onset at different values of r, as well as the varying amplitudes as functions of the trajectory parameters. The equations of motion indicate no effect when either φ' or $\cos\varphi$ is zero. Changing the hyperbola parameters for the actual Pioneer trajectories may change our numerical results, but will not alter the basic model structure. Furthermore, we are not claiming that $\underline{\mathbf{B}}_g$ alone is responsible for all of the observations.

We have not allowed for the effects of $\sigma r' (\cos\varphi)/(2r^2)$ in equation 6b. The term is in fact a small acceleration in the direction of $\underline{\mathbf{u}}_\varphi$, which is perpendicular to $\underline{\mathbf{r}}$, the radius vector from the sun to the spacecraft. To the degree that the path from the Earth to the spacecraft coincides with $\underline{\mathbf{r}}$, this term would not appear directly in the Doppler data, but may appear during the data reduction described above. Without further information we cannot make an assessment, but will leave open the possibility that it may have affected r_c'' as time progressed.

We further point out that we have modeled to a first order level only five of the hyperbolic trajectory parameters for the Pioneers – the periapsis a_p, eccentricity ε, inclination ι, argument of periapsis θ_p, and true anomaly, *i.e.*, the spacecraft position on its trajectory as a function of time. Missing is the independent argument of the trajectory's ascending node Ω relative to the base reference frame. It is Ω which mainly determines the direction of the trajectory following its exit from the solar field.

10 – PRINCIPAL THEORY ASSESSMENTS AND COMPARISONS

In this chapter we will analyze phase invariance for light rays based on its definition and show that experiments that confirm phase invariance do not justify the hyperbolic relationship between space and time propounded by the special theory of relativity. Using a proposed classical analog for the spin states of atomic theory, we compare our gravitational approach with aspects of electromagnetic theory, including direct comparisons between Maxwell's field equations and a corresponding gravitation wave equation.

Phase Invariance And Special Relativity

Maxwell's classical equations encompass the propagation of light waves in empty space, and it was the ability to measure the constants of free space permittivity ε_o and permeability κ_o in a laboratory that led to the acceptance of $c = 1/(\varepsilon_o\kappa_o)^{1/2}$ is a universal constant. Unfortunately, Maxwell accepted the *luminiferous aether* ("light-bearing medium") concept and his theory results were mistakenly used to generate *aether* proposals. Equation 1 is the differential equation in radial coordinates for the wave propagation of an electric field vector $\underline{\mathbf{E}}$ emitted by a point source, where the partial derivatives of $\underline{\mathbf{E}}$ are taken with respect to the r-φ-θ coordinates centered on the light source and t is the time since emission. See Figure 10.2 below.

$$\begin{aligned}\nabla^2\underline{\mathbf{E}} &= (1/r^2)\,\partial/\partial r\,(r^2\,\partial\underline{\mathbf{E}}/\partial r) + (1/r^2\sin\varphi)\,\partial/\partial\varphi\,(\sin\varphi\,\partial\underline{\mathbf{E}}/\partial\varphi) \\ &\qquad + (1/r^2\sin^2\varphi)\,\partial^2\underline{\mathbf{E}}/\partial\theta^2 \\ &= \varepsilon_o\kappa_o\,\partial^2\underline{\mathbf{E}}/\partial t^2 = (1/c^2)\,\partial^2\underline{\mathbf{E}}/\partial t^2, \quad \text{Maxwell's wave equation,} \qquad (1)\end{aligned}$$

$$\begin{aligned}\underline{\mathbf{E}} &= \underline{\mathbf{E}}_o\,(r_o/r)\,\exp i(\underline{\mathbf{k}}\bullet\underline{\mathbf{r}} \pm \omega t), \quad \text{where } \underline{\mathbf{E}}_o \text{ and } r_o \text{ are constants, } i = (-1)^{1/2}, \\ \underline{\mathbf{r}} &= r\,\underline{\hat{\mathbf{u}}}_r,\ \underline{\mathbf{k}} = k\,\underline{\hat{\mathbf{u}}}_r = \omega\,\underline{\hat{\mathbf{u}}}_r/c = 2\pi\,\underline{\hat{\mathbf{u}}}_r/\lambda, \text{ and } \phi = \underline{\mathbf{k}}\bullet\underline{\mathbf{r}} \pm \omega t. \qquad (2)\end{aligned}$$

The vector solution provided by expression 2 for a light ray wave front is radially directed to all distant locations. † Phase is defined as $\phi = \underline{\mathbf{k}}\bullet\underline{\mathbf{r}} \pm \omega t$

---

† Unlike Cartesian coordinates, the radial solution readily specifies a 1/r decrease in the amplitude of $\underline{\mathbf{E}}$ and the geometric extinction of energy proportional to $1/r^2$, as expected.

in expression 15, where $\underline{\mathbf{r}}$ is any spatial location relative to the ray's source at a time t after its arrival, ω is the wave's angular frequency, and c is the speed of light. † The wave front vector $\underline{\mathbf{k}} = k\,\underline{\hat{\mathbf{u}}}_r$ points in the direction of $\underline{\mathbf{r}} = r\,\underline{\hat{\mathbf{u}}}_r$ for a unit vector $\underline{\hat{\mathbf{u}}}_r$ and has a magnitude of $k = \omega/c = 2\pi/\lambda$ for a wave length λ, "exp" is the base of the natural (Napierian) logarithm, $\underline{\mathbf{k}} \bullet \underline{\mathbf{r}} = k\,r$ is the scalar (dot) product of $\underline{\mathbf{k}}$ and $\underline{\mathbf{r}}$, and $\underline{\mathbf{E}}_o$ is a constant vector perpendicular to $\underline{\mathbf{r}}$, where r_o is a constant reference location. A similar expression applies to the magnetic field component $\underline{\mathbf{B}}$, which remains perpendicular to $\underline{\mathbf{r}}$ and $\underline{\mathbf{E}}$.

In experiments conducted during the 1880s at what is now Case Western Reserve University in Cleveland, Ohio, Albert A. Michelson and Edward W. Morley verified that no phase shift occurs for a wave packet as a function of the observer's motion. The expected fringe shift of $2(L/\lambda)\,(v_e^2/c^2)$ due to an aether, where L is the apparatus path length, was not observed.

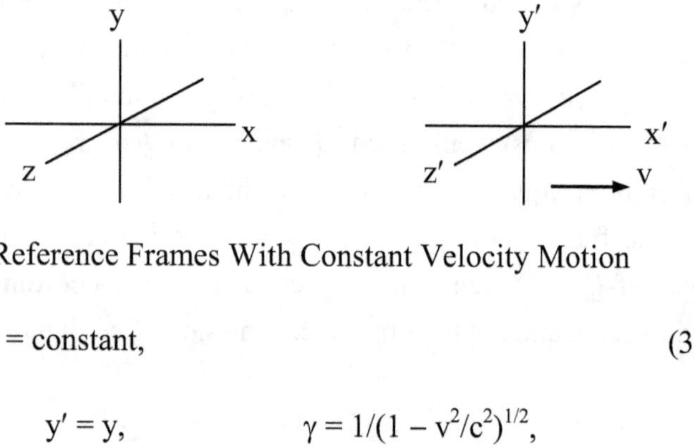

Figure 10.1. Reference Frames With Constant Velocity Motion

$$x^2 + y^2 + z^2 - c^2t^2 = \text{constant}, \qquad (3)$$

$$x' = \gamma\,(x - vt), \qquad y' = y, \qquad \gamma = 1/(1 - v^2/c^2)^{1/2},$$
$$z' = z, \qquad t' = \gamma\,(t - vx/c^2), \qquad \text{special relativity results.} \qquad (4)$$

At the turn of the 20th Century, Albert Einstein built upon the work of Woldemar Voigt, George F. FitzGerald, Joseph Larmor, Hendrik A. Lorentz, and others to develop an invariance theory consistent with the Michelson-Morley results, the constancy of c, and some other experiments. Like its predecessors, Einstein's theory requires the form of expression 3 to remain

---

† See also, Stimson, G.W., *Introduction to Airborne Radar*, Hughes Aircraft Co., El Segundo, CA (1983), at p.87, "Phase is the degree to which the individual cycles of a wave or signal coincide with those of a reference of the same frequency."

constant from reference frame to frame. Anomalously, the Michelson-Morley results have been misinterpreted by substituting $\underline{\mathbf{k}} \bullet \underline{\mathbf{x}}$ for $\underline{\mathbf{k}} \bullet \underline{\mathbf{r}}$ in the phase ϕ of equation 2, where $\underline{\mathbf{x}}$ is the Cartesian coordinate for an observer moving with constant velocity $\underline{\mathbf{v}}_e$ along $\underline{\mathbf{x}}$. See Figure 10.1, where x, y, and z are an object's spatial coordinates at a time t in a given frame, and x′, y′, and z′ are the coordinates at t′ for the same object in the moving frame.

Einstein's purely geometrical theory intermingles spatial position with time, expression 4, but does not specify any physical process whereby the change in the rate kept by a clock should occur. † Its applications involve the Lorentz factor $\gamma = (1 - v^2/c^2)^{-1/2}$, named after Hendrik A. Lorentz (1853-1928), which precludes object velocities from exceeding the speed of light. Since the coordinate transformations are linear, the separation of events #1 and #2 in an unprimed frame, expressed as $\Delta x^2 + \Delta y^2 + \Delta z^2 - c^2\Delta t^2$, where $\Delta x = x_2 - x_1$, etc., obeys the hyperbolic constraint in the primed frame. However, expression 1, and not expression 3, defines phase invariance. ††

The Constancy Of Phase In A Moving Frame

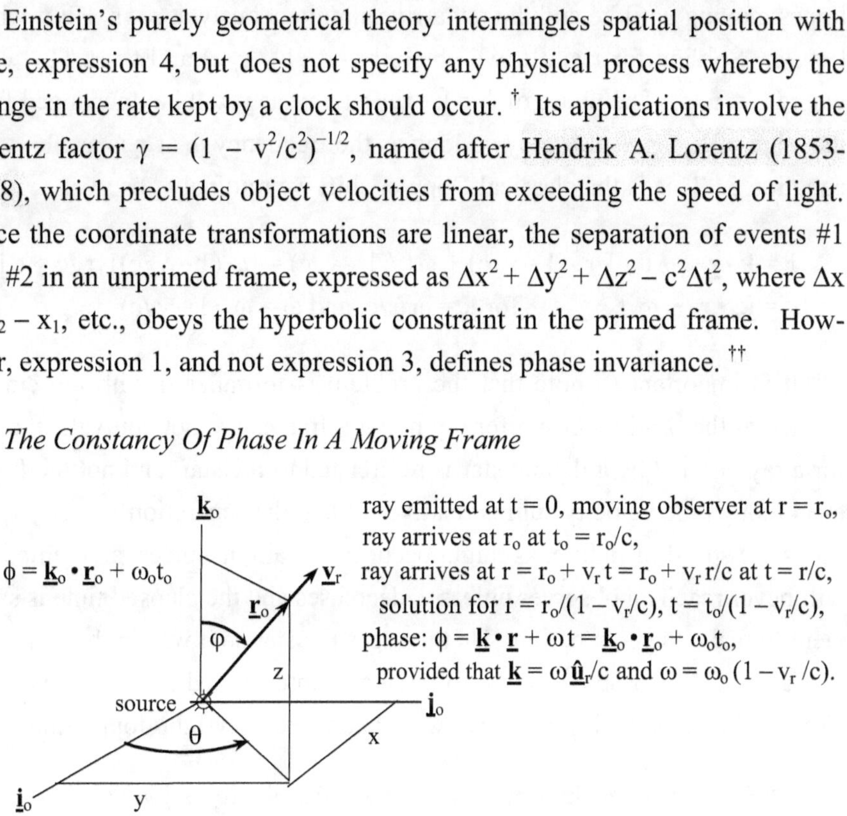

Figure 10.2. Phase Relationships In Spherical Polar Coordinates

---

† In 1899 Lorentz referred to a form of time dilation based on oscillating electron frequencies, which he called "local time", to explain the Michelson-Morley experiment.

†† The Michelson-Morley and Ives-Stilwell experiments have incorrectly been cited as basic tests in support of special relativity. See, Ives, H., Stilwell, G. (1938), "An experimental study of the rate of a moving atomic clock", *Journal of the Optical Society of America*, **28** (7): 215. Ives-Stilwell was designed to detect a *transverse* Doppler component of $1/\gamma$ predicted by the theory, which proved to be non-existent.

Let us consider the geometry depicted in Figure 10.2. An observer at $\underline{\mathbf{r}}_o$ is moving with a constant radial velocity v_r at $t = 0$ when a source at the origin emits a ray of light. The ray arrives at $\underline{\mathbf{r}}_o$ at a time t_o with a frequency of ω_o, a wave number of $\underline{\mathbf{k}}_o = k_o\,\underline{\hat{\mathbf{u}}}_r = \omega_o\,\underline{\hat{\mathbf{u}}}_r/c$, and a phase of $\phi = \underline{\mathbf{k}}_o \bullet \underline{\mathbf{r}}_o + \omega_o t_o$. Although $\underline{\mathbf{r}}$ and t vary separately, t_o must be greater than or equal to r_o/c in order for the ray to be detected at $\underline{\mathbf{r}}_o$. At t_o the observer has moved and continues moving until he and the ray arrive simultaneously at $r = r_o + v_r t$, where $t = r/c$. Solving for r and t, we obtain $r = r_o/(1 - v_r/c)$ and $t = t_o/(1 - v_r/c)$, where $t_o = r_o/c$. The phase of $\phi = \underline{\mathbf{k}} \bullet \underline{\mathbf{r}} + \omega t + \phi_o$ at $\underline{\mathbf{r}}$ will be unchanged from its value of $\underline{\mathbf{k}}_o \bullet \underline{\mathbf{r}}_o + \omega_o t_o$ at $\underline{\mathbf{r}}_o$ so long as the frequency ω at $\underline{\mathbf{r}}$ is $\omega_o\,(1 - v_r/c)$, equation 5. This is the classical Doppler shift we routinely observe.

$$\phi = \underline{\mathbf{k}} \bullet \underline{\mathbf{r}} + \omega t = k r_o/(1 - v_r/c) + \omega t_o/(1 - v_r/c) = [\omega/(1 - v_r/c)]\,[r_o/c + t_o]$$
$$= \underline{\mathbf{k}}_o \bullet \underline{\mathbf{r}}_o + \omega_o t_o, \qquad \text{for } \underline{\mathbf{k}} = \omega\,\underline{\hat{\mathbf{u}}}_r/c, \text{ and } \omega = \omega_o\,(1 - v_r/c). \qquad (5)$$

It is important to note that the problem is formulated with the emitter located at the origin. All uniformly moving frames are not equivalent, and $\underline{\mathbf{r}}$ for a ray emitted by a distant star is referenced to the star, and not the Earth, even though the velocity sum is relative. When the separation velocity v_r exceeds the speed of light c, as might occur in certain instances, an emitted ray will never reach an observer unless v_r decreases and the elapsed time is sufficient to make the connection. He then sees the star as it was in the past. We will address this problem in the following chapter, based on universe expansion. But compare with, http://www.astro.ucla.edu/~wright/doppler.htm,

> "[F]or the largest known redshift of z = 6.3, the recession velocity is not 6.3 c = 1,890,000 kilometers per second. It is also not the 285,254 kilometers per second given by the special relativistic Doppler formula $1 + z = [(1 + v/c)/(1 - v/c)]^{1/2}$. The actual recession velocity for this object depends on the cosmological parameters, but for an Omega_M = 0.3 vacuum-dominated flat model the velocity is 585,611 kilometers per second. This is faster than light." †

† See also, https://arxiv.org/pdf/astro-ph/0011070v2.pdf, discussing the Omega_M = 0.3 flat model and proposing "two forms of velocity". We note that the observation of $z = (\lambda_{obs} - \lambda_o)/\lambda_o = 6.3$ occurs near the current limits of luminosity detectability.

Au contraire. The classical Doppler formula for a stationary source and a moving observer is $\omega = \omega_o(1 - v_r/c)$, and not $\omega = \omega_o/(1 + v_r/c)$. The special theory achieves phase constancy by multiplying $\omega = \omega_o/(1 + v_r/c)$ by $1/\gamma$ so that quadratic expression 16 remains constant from frame to frame. Effects of γ are ignored for velocities less than c, and experiments † claimed to validate the presence of $1/\gamma$ are based on null results and circuitous assumptions. The expressions $r = r_o/(1 - v_r/c)$ and $t = t_o/(1 - v_r/c)$ are communication constraints for a moving observer and do not imply that space has expanded or that lapses of time have mysteriously lengthened, as advocated by relativists.

The (Non Parallex) Aberration Of Starlight

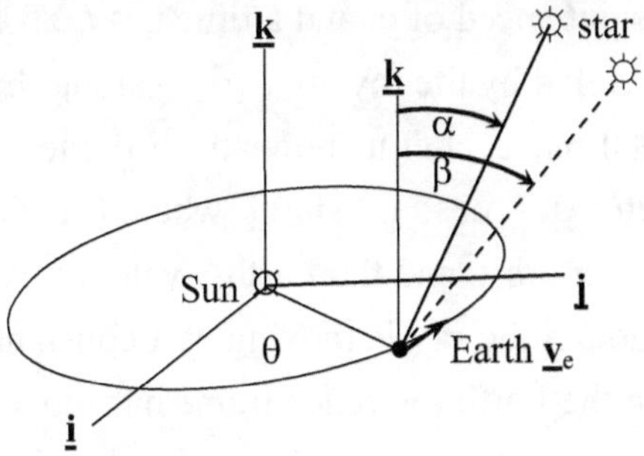

Figure 10.3. Starlight Aberration Geometry (Exaggerated)

The proper modeling of a phenomenon called the (non-parallax) aberration of starlight supports our approach and shows the special theory to be superfluous for its explanation. The aberration has been observed since the late 1600s and involves an apparent shift in the view of a star's position by as much as about ±20.5 arc-seconds during any year. In Figure 10.3, β is the viewing angle observed at any location in the moving Earth's orbital plane, $\underline{v}_e$ (of magnitude v_e) is the Earth's velocity and α is the angle observed by a

---

† Since the relativistic shift is $\omega = \omega_o (1 - v^2/c^2)^{1/2}$ for purely non-radial velocities v, various adjustments have been made in futile attempts to verify an alleged transverse Doppler shift. The Kennedy-Thorndike experiment, purporting to verify the presence of $1/\gamma$ for time transformations, assumed that the factor was already present in the form of a length modification. See, Kennedy, R. and Thorndike, E., (1932), "Experimental Establishment of the Relativity of Time", *Physical Review*, **42** (3): 400–418.

stationary frame at the same location. It was first explained (incorrectly) by James Bradley in 1727 † by adding the Earth's velocity v_e at various orbit locations to the speed of light c. Bradley's approach leads to the result $\cot\beta = (\cos\alpha)/(\sin\alpha + v_e/c)$. For v_e much less than c, the result becomes $\tan(\alpha - \beta) \cong \alpha - \beta = -v_e/c$ at $\alpha = 0$ degrees. However, Bradley's theory was incompatible with other theories of light at the time and later, when analysts were focused on the *luminiferous aether* for the propagation of light.

Let us instead imagine a stationary telescope of length l aligned with the solid line pointed toward the star in the **jk** plane of Figure 10.3, and a star so far away that angle α does not vary within our measurement capability for a telescope located at any position on the Earth's orbit. A light ray travels down the telescope at a constant speed of c, and a time $t_l = l/c$ is required for its transit. The distances traveled by the ray are $l\cos\alpha$ along the **k** axis and $l\sin\alpha$ in the Earth's **ij** orbital plane. Let us now attach the telescope to the Earth's moving position vector $\underline{\hat{\mathbf{u}}}_e = \cos\theta\,\underline{\mathbf{i}} + \sin\theta\,\underline{\mathbf{j}}$, where θ is the azimuthal angle for a nearly circular Earth orbit, and the Earth's velocity is given by $\underline{\mathbf{v}}_e = v_e(-\sin\theta\,\underline{\mathbf{i}} + \cos\theta\,\underline{\mathbf{j}})$. Although the ray is moving at a constant velocity c, the telescope's focal plane in the Earth reference frame has moved a distance of $r_{ij} = v_e t_l = v_e l/c$ in the **ij**-plane during the ray's transit, as determined by θ.

Forming the two position vectors $\underline{\boldsymbol{l}}_1$ and $\underline{\boldsymbol{l}}_2$ of lengths of l_1 and l_2, expressions 6a and 6b, we may model the distance traveled by the light ray for the inertial path as l_1, with l_2 representing the path that includes the moving Earth. From 6b we see that all of the changes of $\underline{\boldsymbol{l}}_2$ from $\underline{\boldsymbol{l}}_1$ are of order v_e/c and lie in the **ij**-plane. Using the dot product $\underline{\boldsymbol{l}}_1 \bullet \underline{\boldsymbol{l}}_2 = l_1 l_2 \cos\delta$, where δ is the angle between the two position vectors, we may express δ by equation 6c.

$$\underline{\boldsymbol{l}}_1 = l\,(\sin\alpha\,\underline{\mathbf{j}} + \cos\alpha\,\underline{\mathbf{k}}), \qquad l_1 = l, \tag{6a}$$

$$\underline{\boldsymbol{l}}_2 = l\,\{[(-v_e\sin\theta)/c]\,\underline{\mathbf{i}} + [\sin\alpha + (v_e\cos\theta)/c]\,\underline{\mathbf{j}} + \cos\alpha\,\underline{\mathbf{k}}\},$$
$$l_2 = l\,[1 + 2v_e(\sin\alpha\cos\theta)/c + v_e^2/c^2]^{1/2}, \tag{6b}$$

---

† See, Bradley, J. (1727–1728), "A Letter from the Rev. Mr. James Bradley, Savilian Professor of Astronomy at Oxford, and F.R.S. to Dr. Edmond Halley Astronom. Reg. & c. Giving an Account of a New Discovered Motion of the Fix'd Stars", *Phil. Trans. R. Soc.* **35**: 637. The work provided the first credible estimate of the speed of light.

$$\cos\delta = [1 + v_e(\sin\alpha\cos\theta)/c]/[1 + 2v_e(\sin\alpha\cos\theta)/c + v_e^2/c^2]^{1/2}, \qquad (6c)$$

$$\text{At } \alpha = 0, \quad \underline{\boldsymbol{l}}_1 = l\,\underline{\mathbf{k}}, \quad \cos\delta = 1/(1 + v_e^2/c^2)^{1/2}, \ \tan\delta = v/c, \text{ for all } \theta, \qquad (6d)$$

$$\text{At } \alpha = \pi/2, \ \underline{\boldsymbol{l}}_1 = l\,\underline{\mathbf{i}},$$
$$\cos\delta = [1 + (v_e\cos\theta)/c]/[1 + 2v_e(\cos\theta)/c + v_e^2/c^2]^{1/2}. \qquad (6e)$$

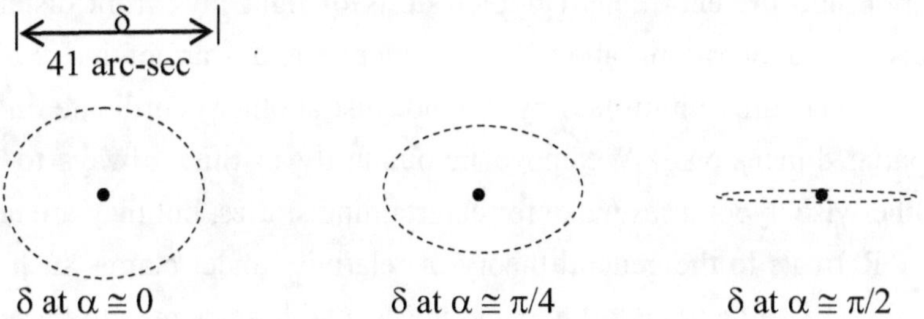

Figure 10.4. Results For Annual Aberration Of Starlight

From expressions 6b and 6d we see that for $\alpha = 0$ (looking straight up) the magnitude of δ is constant at $v_e/c \cong 10^{-4}$ radian $\cong$ 20.5 arc-seconds, as its direction rotates with θ in the **ij**-plane. See, Figure 10.4, first sketch. Increasing α to $\pi/2$, *i.e.*, placing the line of sight in the **ij**-plane, equation 6e shows that δ ranges from 0 to $\pm$ 20.5 arc-seconds as θ varies. *Ibid.*, last sketch. At intermediate values of α the circular form for the rotation of δ at $\alpha = 0$ gradually flattens to that for $\alpha = \pi/2$ as the viewing aspect angle varies. This is the observed form of starlight aberration.

Contraindications Of The Special Theory

Our derivation of starlight aberration depends on correctly modeling the effect of v^2/c^2 and will not accommodate an additional input of the Lorentz factor $\gamma = 1/(1 - v^2/c^2)^{1/2}$ into the result. We also decline to accept relativity arguments [†] that 4×4 matrix expressions for electromagnetic field components resolve a conflict between time-independent **B** forces in a given system versus moving ones. The velocity of an object relative to a field source can

† Cf., Einstein, A. (1905), "On the Electrodynamics of Moving Bodies", Eng. Tr., http://www.fourmilab.ch/etexts/einstein/specrel/www/.

be detected by **B**, a magnetic type of force, and the *Lorentz condition* is not satisfied for a static vector potential whose divergence is not zero. It again follows that all frames moving with constant velocities are not equivalent. A specific set of coordinates is best suited to describe a field in a given region, even though the concept of an *aether-like* "fabric of space-time" is rejected.

Issues involving misapplication of the special theory to starlight aberration and the enormous Doppler shifts of light rays from distant galaxies raise valid questions about its acceptance and warrant its reexamination. Moreover, time multiplied by c is not just another coordinate on a par with spatial dimensions. We move involuntarily in time, always to the future. Otherwise hypotheses make for entertaining stories, but they are not reality.

Retreats to the general theory of relativity under claims such as the ambiguity of velocity are ad hoc and unacceptable responses to these problems. Notwithstanding never-say-die efforts by theorists to validate the general theory, the very small perturbations upon which it ultimately rests are non-exotic phenomena and have other reasonable explanations. [†] One example is the perihelion advance of Mercury's orbit, [††] as discussed above. Similarly, red shifts for light rays in the Earth's gravitational field can be modeled by perturbations to the equations of atomic physics. Other general theory computations, including the deflection of light rays near the sun, rely on a Schwarzschild radius r_s which, notwithstanding involved plausibility arguments, is *a freely adjustable constant of integration.* [†††] Although it may be logically sound, a physics theory must be tested by its ability to model observations, and if it fails, it must be rejected. Observations of future spacecraft orbits should resolve issues of the theory's applicability to the solar system.

[†] Physicists have gone to great lengths to accommodate relativity theory. One example is the bias in interpreting the observations of light ray bending by the sun during a 1919 solar eclipse, which at least indicates a dependence on viewing geometry.

[††] The general theory predicts perihelion advances for Venus of 8.6 arc-seconds per century and 3.8 arc-seconds per century for Earth, which we have not seen validated.

[†††] See, Stephani, *General Relativity*, *supra*, p.116. The author casually dismisses null results of more precise measurements provided by artificial Earth satellites.

The Electron Spin Hypothesis Of Atomic Theory

The Schroedinger equation of atomic theory uses the reduced Planck (angular momentum) constant $\hbar = h/(2\pi)$ to model electron orbits analogous to orbits in gravitational fields. The nuclear attraction is based on an electric field $\underline{\mathbf{E}} = -Ze\, \underline{\hat{\mathbf{u}}}_r/(4\pi\varepsilon_o r^2)$, where Z is the atomic number of the nucleus, e is the charge of the electron and proton, ε_o is the permittivity of free space, r is the orbit radius $|\underline{\mathbf{r}}|$, and $\underline{\hat{\mathbf{u}}}_r$ is a unit vector in the direction of $\underline{\mathbf{r}}$. During its early stages the theory failed to account for observed angular momentum states involving half-integral values of $\hbar$ for some atoms, especially alkali metals. The dilemma led to the hypothesis that the electron possesses *intrinsic* angular momentum in the form of electron spin. It is represented by a vector $\underline{\mathbf{s}}_e$, specifying a magnetic moment of $\underline{\boldsymbol{M}}_e = -e\underline{\mathbf{s}}_e/m_e$, where m_e is the electron's mass and $e/(2m_e)$ is called the Bohr magneton. The square of the spin magnitude is $\underline{\mathbf{s}}_e \bullet \underline{\mathbf{s}}_e = s_e\,(s_e + 1)\,\hbar^2$, where $s_e = 1/2$, and the projection of spin on the atom's polar axis is taken to be $\pm\,\hbar/2$, *i.e.*, spin up or down. †

Relying on matrix operators to model the electron's energy, P.A.M. Dirac developed a theory to support the ad hoc assumptions. He applied the Lorentz transformation of special relativity to model a magnetic field of $\underline{\mathbf{B}} = -\underline{\mathbf{v}} \times \underline{\mathbf{E}}/c^2$ in a reference frame for the electron, as if it were moving with a constant velocity of $\underline{\mathbf{v}}$ rather than being in an orbit. Dirac's theory is based on the hypothesis that the electron magnetic moment interacts with the magnetic field that *its own motion* has created to produce an energy term $\underline{\boldsymbol{M}}_e \bullet \underline{\mathbf{B}} = [Ze^2/(4\pi\varepsilon_o c^2)]\,[\underline{\mathbf{s}}_e \bullet (\underline{\hat{\mathbf{u}}}_r \times \underline{\mathbf{v}})/(2m_e r^2)]$, where c is the speed of light.

The spin theory begs the question of why spin interactions are not observed for all atoms since identical electrons are continually moving through the nuclear electric field and the effect should always exist. There is also a logic disconnect in first viewing the field from the nuclear perspective and then adding an adjustment based on the electron's non-inertial reference frame. But the theory's most troubling aspect is that the spin interaction has no classical counterpart, in violation of the correspondence principle. The

† The approach is comparable to implausibly assigning spin angular momentum to the Earth in an amount of about five percent of the Earth's *orbital* angular momentum.

electron may indeed possess a form of spin, similar to spinning planets, but we dispute the claim that half-integral *orbital* energy states can be attributed to its spin interaction. They are instead due to the interaction of the curl of a nuclear vector potential, $\underline{\mathbf{B}} = \nabla \times \underline{\mathbf{A}}$, with the current loop produced by the electron's motion. † Its occasional appearance indicates that only certain nuclei possess $\underline{\mathbf{A}}$, with orbital energy of the form provided by our theory.

Restatement Of Electron Spin Theory

Our theory would treat the electron's motion as a current of –eW for an orbital angular frequency W. The current produces a magnetic moment of $\underline{\boldsymbol{M}}_e = -(eW)(4r^2)$ and a torque of $\underline{\boldsymbol{\tau}} = N\, \underline{\boldsymbol{M}}_g\, \underline{\mathbf{B}}_{gav}$, where $\underline{\mathbf{B}}_{gav}$ is the orbital average of the magnetic field produced by the nucleus, and N is the orbit's multiple of the Planck constant ħ. The body frame of the electron is inclined to the polar axis of the atom, and frame rotation is occurring so that $m_e^2\, \underline{\mathbf{J}}_{orb}^{\ 2} = [j\,(j+1) - s\,(s+1)]\,\hbar^2$, where $j = m - s$ and m is the azimuth constant for the wave solution. This is the observed form for $s = +1/2$, where the total angular momentum is $m_e\, \underline{\mathbf{J}}_{oz} = [j\,(j+1)]^{1/2}\, \hbar\, \underline{\mathbf{k}}_o$, including frame rotation.

If we replace our angular momentum parameter ϭ by $\hbar/m_e$ and the solar gravitational force constant $\mu_p M g_o$ by the atomic force constant $Ze^2/(4\pi\varepsilon_o)$, with the nuclear magnetic field scaling in proportion to Z, equation 7 specifies a vector potential for unit vectors of $\underline{\hat{\mathbf{u}}}_r$, $\underline{\hat{\mathbf{u}}}_\varphi$, and $\underline{\hat{\mathbf{u}}}_\theta$, where $e\underline{\mathbf{B}}$ is given by expression 8. The end forms are obtained by using the *fine structure constant* $\alpha_e = e^2/(4\pi\varepsilon_o\hbar c) \cong 1/137.04$ and expressing $1/(\varepsilon_o c^2)$ as free-space magnetic permeability $\kappa_o = 4\pi \times 10^{-7}$ kilogram meters per coulomb squared.

$$e\underline{\mathbf{A}} = Z\hbar\,(\sin\varphi\, \underline{\hat{\mathbf{u}}}_r + \cos\varphi\, \underline{\hat{\mathbf{u}}}_\varphi)/(2r)$$
$$= [Ze^2/(4\pi\varepsilon_o c\alpha_e)]\,(\sin\varphi\, \underline{\hat{\mathbf{u}}}_r + \cos\varphi\, \underline{\hat{\mathbf{u}}}_\varphi)/(2r), \qquad (7)$$

$$e\underline{\mathbf{B}} = \nabla \times e\underline{\mathbf{A}} = -\,[Ze^2/(4\pi\varepsilon_o c\alpha_e)]\,(\cos\varphi)\, \underline{\hat{\mathbf{u}}}_\theta/(2r^2)$$
$$= -\,(c/\alpha_e)\,[Ze^2\kappa_o/(4\pi)]\,(\cos\varphi)\, \underline{\hat{\mathbf{u}}}_\theta/(2r^2). \qquad (8)$$

---

† The Stern-Gerlach experiment, wherein silver atoms are projected through an inhomogeneous magnetic field, is consistent with a nuclear vector potential.

In Chapter 4 we proposed a magnetic-like flux $\underline{\mathbf{B}}_g$ for the solar gravitational field of the type indicated in expression 9. If we replace g_o by $1/(4\pi\varepsilon_g)$ and define c_g, $\mu_p M\kappa_g$, and α_g as the counterparts of c, $Ze^2\kappa_o$, and α_e in expression 8 with $1/c_g^{\,2} = \varepsilon_g\kappa_g$, the atomic field analogy provides the second line of expression 9. Based on a fit of the solar data, we have obtained a value for $\sigma\!\!\!-$ and thus for the product of the terms contained in expression 10.

$$\begin{aligned}\mu_p\underline{\mathbf{B}}_g = \nabla\times\mu_p\underline{\mathbf{A}}_g &= -\,\mu_p\sigma\!\!\!-\,(\cos\varphi)\,\underline{\hat{\mathbf{u}}}_\theta/(2r^2)\\ &= -\,(c_g/\alpha_g)\,[\mu_p M\kappa_g/(4\pi)]\,(\cos\varphi)\,\underline{\hat{\mathbf{u}}}_\theta/(2r^2),\end{aligned} \qquad (9)$$

$$\begin{aligned}\sigma\!\!\!- &= (c_g/\alpha_g)\,[M\kappa_g/(4\pi)] = (c_g/\alpha_g)\,[M/(4\pi\varepsilon_g c_g^{\,2})]\\ &= Mg_o/(\alpha_g c_g) \cong 1.946\times10^9 \text{ kilometers squared per second,}\end{aligned} \qquad (10)$$

$$\alpha_g c_g = Mg_o/\sigma\!\!\!- = (Mg_o/a_o)^{1/2} \cong 68.25 \text{ kilometers per second,}$$
$$\text{where } a_o \cong 0.1905 \text{ astronomical units,} \qquad (11)$$

$$(Mg_o/\sigma\!\!\!-)/\alpha_g = c_g = c \cong 2.998\times10^8 \text{ meters per second,}$$
$$\text{so that } \alpha_g \cong 1/4393. \qquad (12)$$

When we multiply the solar mass $M \cong 1.991\times10^{30}$ kilograms by $g_o \cong 6.67\times10^{-20}$ kilometers cubed per second squared per kilogram, we obtain $Mg_o \cong 1.328\times10^{11}$ kilometers cubed per second squared. Thus, $\alpha_g c_g$ in expression 11 must be about 68.25 kilometers per second, which is also the value of $(Mg_o/a_o)^{1/2}$ for a_o estimated at 0.2850×10^7 kilometers $\cong$ 0.1905 astronomical units. The atomic value is $\alpha_e c \cong c/137.04 \cong 2188$ kilometers per second. Since Mercury's perihelion advancement indicates that $c_g = c$, we obtain $1/\alpha_g \cong 4393$, expression 12, for gravity's fine structure constant.

The Search For A Unified Theory

Our approach provides a unified gravity theory not previously addressed in available literature. Instead of forcing the same parameters on the vastly different atomic and gravitational fields, we accept that each propagates its specific properties, but require the gravitational field components to exhibit unification comparable to the electric and magnetic fields, as formulated by

James Clerk Maxwell in 1873 without any deference to gravity.[†] The occurrence of velocity-dependent forces is an inherent property of classical fields, and the force occurs even when the current elements are identical.

Such an approach for gravity was never undertaken, as the physics community first seemed to be content with the simple inverse square attraction formulated by Isaac Newton some 300 years ago. Then, at the turn of the 20th century, physicists "left the reservation" and eagerly embraced the metaphysics of relativity, ignoring the evidence of a magnetic-like force in the solar gravitational field. The magnetic-like component has been apparent in the solar field since the Titius-Bode "law" and planetary inclinations were first observed, but the community has not been sufficiently clever to discern it. Rather than proposing an electromagnetic analogy and relying on Euler's methods, general relativists have pursued four-dimensional, non-linear geometry in complicated attempts to explain observations. Equally distressing, when faced with anomalies in the trajectories of interplanetary spacecraft, analysts rejected "new physics" out of hand and instead embraced an ad hoc heat radiation theory which fails for spacecraft other than the Pioneers.

In order to clarify the electromagnetic analogy, let us define the static gravitational field component $\underline{\mathbf{G}}$ in free space as Newton's force $\underline{\mathbf{F}}_g$ divided by the mass μ of a test object in the field, *i.e.*, $\underline{\mathbf{G}} = \underline{\mathbf{F}}_g/\mu$. Replacing g_o by $1/(4\pi\varepsilon_g)$, $\underline{\mathbf{G}}$ takes the form of equation 13b, where M is the mass of the field source. $\underline{\mathbf{G}}$ mirrors the electric field $\underline{\mathbf{E}} = \underline{\mathbf{F}}_g/e$, equation 13a, where e is the electron charge, Z is the atomic number, and ε_o is free-space permittivity.

$$\underline{\mathbf{E}} = -Ze\,\hat{\underline{\mathbf{u}}}_r/(4\pi\varepsilon_o r^2) = [Ze/(4\pi\varepsilon_o)]\,\nabla(1/r), \qquad (13a)$$

$$\underline{\mathbf{G}} = -M\,\hat{\underline{\mathbf{u}}}_r/(4\pi\varepsilon_g r^2) = [M/(4\pi\varepsilon_g)]\,\nabla(1/r), \qquad (13b)$$

$$\nabla \bullet \underline{\mathbf{E}} = \rho/\varepsilon_o, \qquad \nabla \bullet \underline{\mathbf{G}} = \rho_g/\varepsilon_g, \qquad (14)$$

$$\nabla \bullet \underline{\mathbf{B}} = 0, \qquad \nabla \bullet \underline{\mathbf{B}}_g = 0, \qquad (15)$$

[†] Maxwell based portions of his famous equations on the experimental work of Michael Faraday (1791-1867), a great pioneer in research physics who did not possess the educational background needed to generate the formal equations.

$$\nabla \times \underline{\mathbf{E}} = -\,\partial \underline{\mathbf{B}}/\partial t, \qquad \nabla \times \underline{\mathbf{G}} = -\,\partial \underline{\mathbf{B}}_g/\partial t, \tag{16}$$

$$\nabla \times \underline{\mathbf{B}} = (1/c^2)\,\partial \underline{\mathbf{E}}/\partial t, \qquad \nabla \times \underline{\mathbf{B}}_g = (1/c_g^{\,2})\,\partial \underline{\mathbf{G}}/\partial t, \tag{17}$$

$$\oint \underline{\mathbf{E}} \bullet d\underline{l} = \int \nabla \times \underline{\mathbf{E}} \bullet d\underline{\mathbf{S}} = -\,\partial/\partial t \int \underline{\mathbf{B}} \bullet d\underline{\mathbf{S}}, \tag{18a}$$

$$\oint \underline{\mathbf{G}} \bullet d\underline{l} = \int \nabla \times \underline{\mathbf{G}} \bullet d\underline{\mathbf{S}} = -\,\partial/\partial t \int \underline{\mathbf{B}}_g \bullet d\underline{\mathbf{S}}. \tag{18b}$$

Assuming that there are no dielectric-type or magnetic-type of moments external to the field source, and replacing the mass M by a mass density ρ_g integrated over the source volume, we offer the **G** portions of equations 14, 15, 16, and 17 for the gravitational equivalents of Maxwell's free-space field equations, where $\underline{\mathbf{B}}_g$ is the gravitational equivalent of magnetic flux **B**. Equation 14 is derived by applying the divergence theorem to equation 13, and equation 15 follows from the identity $\nabla \bullet \underline{\mathbf{B}}_g = \nabla \bullet \nabla \times \underline{\mathbf{A}}_g \equiv 0$. Equation 16 should hold for $\underline{\mathbf{B}}_g$ when it is changing in time t. The equation involves Stokes' theorem, as in equation 18b, where **S** is a surface through which $\underline{\mathbf{B}}_g$ is flowing and $\underline{l}$ is the closed path around the surface's perimeter. The end form of equation 18a for **E** is observed in the operation of a generator, for example, when wire rings "cut through" magnetic flux **B** "lines of force".

Computing $\nabla \times (\nabla \times \underline{\mathbf{E}})$ from equation 10 and substituting from equation 16, we obtain equation 19a. But since $\nabla \times (\nabla \times \underline{\mathbf{E}}) \equiv \nabla (\nabla \bullet \underline{\mathbf{E}}) - \nabla^2 \underline{\mathbf{E}}$ for any vector, and since $\nabla \bullet \underline{\mathbf{E}} = 0$ in the space external to the source, we obtain the wave equation $\nabla^2 \underline{\mathbf{E}} = \varepsilon_o \kappa_o\, \partial^2 \underline{\mathbf{E}}/\partial t^2$, where $\varepsilon_o \kappa_o = 1/c^2$. Although we do not have solar data to verify the temporal behavior specified by equations 16 and 17, the potential exists by observing cosmic events, and we expect that $\nabla^2 \underline{\mathbf{G}} = \varepsilon_g \kappa_g\, \partial^2 \underline{\mathbf{G}}/\partial t^2 = (1/c_g^{\,2})\, \partial^2 \underline{\mathbf{G}}/\partial t^2$, equation 19b. Observable gravity waves traveling at the speed c_g should be generated when **G** is changing in time t.

$$\begin{aligned} \nabla \times (\nabla \times \underline{\mathbf{E}}) &= -\nabla \times \partial \underline{\mathbf{B}}/\partial t = -\,\partial(\nabla \times \underline{\mathbf{B}})/\partial t = -\,\varepsilon_o \kappa_o\, \partial^2 \underline{\mathbf{E}}/\partial t^2 \\ &\equiv \nabla (\nabla \bullet \underline{\mathbf{E}}) - \nabla^2 \underline{\mathbf{E}} = -\,\nabla^2 \underline{\mathbf{E}} = -\,\varepsilon_o \kappa_o\, \partial^2 \underline{\mathbf{E}}/\partial t^2, \end{aligned} \tag{19a}$$

$$\nabla \times (\nabla \times \underline{\mathbf{G}}) = -\,\nabla^2 \underline{\mathbf{G}} = -\,\varepsilon_g \kappa_g\, \partial^2 \underline{\mathbf{G}}/\partial t^2. \tag{19b}$$

11 – A STRAIGHTFORWARD APPROACH TO COSMOLOGY

Cosmology theories must account for velocity-to-distance relationships observed for the galaxies, † $v_n = H_o r_n$, where r_n is the distance from the Earth to the nth galaxy, v_n is its velocity, and H_o is the *Hubble constant* named after astronomer Edwin P. Hubble (1889-1953). Doppler data for distant galaxies do not agree with relativity theory, and advocates claim that the "velocity-to-distance relationship of Hubble's law should be viewed as a theoretical result with velocity to be connected with observed redshift, not by the Doppler effect, but by a cosmological model relating recessional velocity to the expansion of the Universe." †† A universe expansion model is indeed needed, but it must correctly incorporate Hubble's law and observed Doppler shifts.

Figure 11.1 depicts a set of observations indicating $H_o \cong 49.2$ kilometers per second per mega-parsec (MPc) $\cong 5.02 \times 10^{-11}$ per year. ††† The age of the universe was estimated at 13.3 billion years by multiplying $1/H_o$ by 2/3.

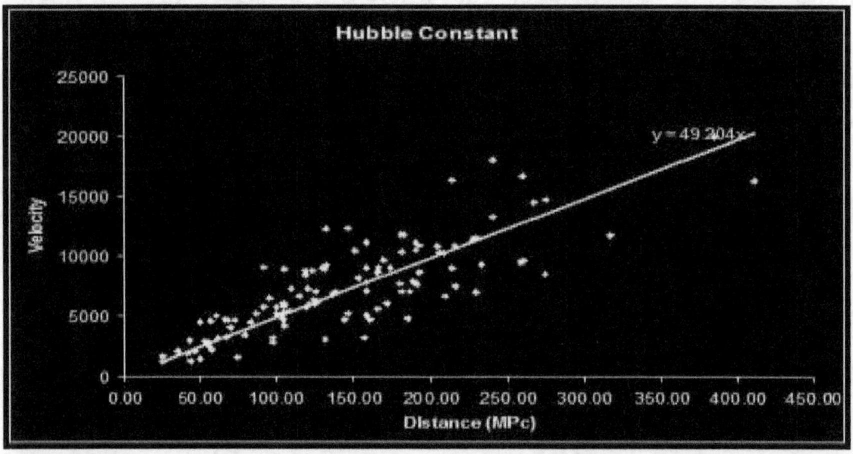

Figure 11.1. Recession Velocities (km/sec) Vs. Galaxy Distances (MPc)
1 MPc $\cong 3.09 \times 10^{19}$ kilometers $\cong$ 3.26 million light-years (Credit: ASNSW)

---

† The observable universe is estimated to contain many billions of galaxies.

†† Harrison, E. (1992), "The Redshift-Distance and Velocity-Distance Laws", *The Astrophysical Journal*, **403**: 28–31.

††† See, http://www.asnsw.com/node/720, based on supernova data from 2000-2003.

Data Acquisition Considerations

Before interpreting the observations, let us briefly consider how astronomical distances are determined. For nearby stars the r_n distances from the Earth are calculated by using the Earth's orbit diameter as a baseline and observing the subtended inclination angles at six months intervals. This process is called *parallax*, and (thanks to modern technologies) works for stars as far away from the Earth as about 15,000 light-years. At greater distances the inclination angle differences over the Earth's orbit are too small to discern with most current capabilities. However, astronomers are able to locate within more distant clusters certain stars called *Cepheid variables*, whose emissions vary in distinctive patterns and have known absolute brightness. By observing their apparent brightness, the intensity reductions versus distance provide estimates of r_n. For distances beyond about 130 million light-years, or 40 mega parsecs (the lower left corner of Figure 11.1), emissions of the *Cepheids* are too faint to support determinations of r_n.

Nevertheless, r_n can be estimated from the assumed absolute brightness of very distant, randomly-occurring supernovas which briefly outshine their entire galaxies. For all of the computations, galaxy velocities relative to the Earth are based on Doppler shifts in the spectra of their emitted rays. †

The Hubble relationship, $v_n = \partial r_n/\partial t = H_o r_n$, seems to indicate a solution of $r_n = r_{no} \exp(H_o t)$, where r_n is the distance from the nth galaxy to the Earth at time t and r_{no} is the present distance. †† However, the form is based on a partial derivative, and the proposed solution is inconsistent with a big bang. For example, locating the Milky Way at r = 0 some 13.8 billion years ago would place other galaxies millions of light-years away when the bang occurred.

---

† Since the maximum velocity plotted in Figure 11.1 is less than 7 percent of the speed of light, we will not address *proper time* as espoused by special relativity.

†† The mean distance between galaxies is roughly 100 times their average diameter. The Milky Way diameter, for example, ranges from about 100,000 to 180,000 light-years. Galaxies tend to occur in clusters, and some of the clusters form superclusters, spanning as much as 200 million light-years. Superclusters are the largest known structures in the universe. See, https://www.britannica.com/topic/supercluster.

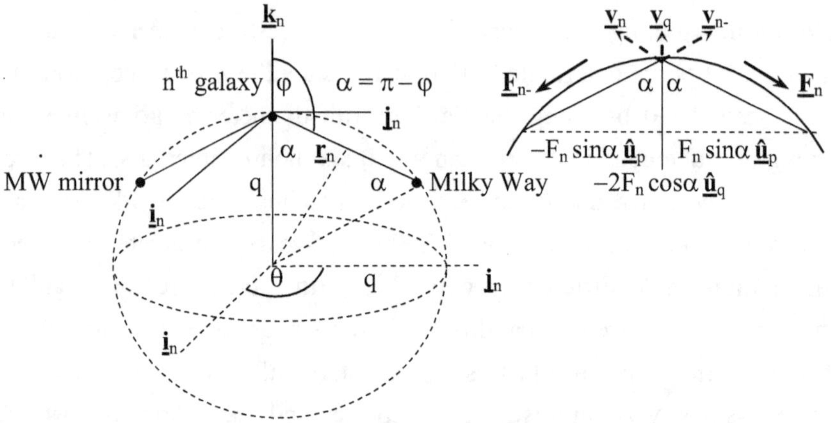

Figure 11.2. Ideal Isotropic Expansion Of The Universe

Figure 11.2 depicts an isotropic, symmetric distribution of galaxies relative to an nth galaxy, which is located at $\varphi = \pi/2$ at all times greater than t_o following a big bang at $t = 0$. The galaxies lie on the surface of a sphere of radius q, and are evenly spaced with a surface density of $\rho = M_t/(4\pi q^2)$ for a total universe mass of M_t. The Milky Way mass M_m is located at $\mathbf{r}_n$ in the $\mathbf{i}_n\mathbf{k}_n$ plane of the nth galaxy, whose own mass is M_n. Many other galaxies are located at the same value of φ as θ ranges from 0 to 2π, and the mass sum at φ can be expressed as $\rho\, dA = -2M_t \sin\varphi \cos\varphi\, d\varphi$ for $\pi/2 \le \varphi \le \pi$. A perpendicular line drawn to $\mathbf{r}_n$ from the origin shows that $r_n/2 = q \cos\alpha = -q \cos\varphi$, where $\varphi = \pi - \alpha$ is the constant declination of the ith galaxy relative to M_n.

The radius of the sphere is expanding continually as $q = \int v_q\, dt$, where the expansion velocity $\mathbf{v}_q = v_q\, \hat{\mathbf{u}}_q$ is normal to the surface at all locations and decreases with time due to gravitational forces of $\mathbf{F}_n$. The recession velocity $\mathbf{v}_n = v_n\, \hat{\mathbf{u}}_r$ between galaxies lies along $\mathbf{r}_n = r_n\, \hat{\mathbf{u}}_r$, where $\hat{\mathbf{u}}_r$ and $\hat{\mathbf{u}}_q$ are unit vectors and $\hat{\mathbf{u}}_r$ is inclined to $\hat{\mathbf{u}}_q$ at the time independent angle α. The force between the galaxies is given by $F_n (\cos\alpha\, \hat{\mathbf{u}}_q + \sin\alpha\, \hat{\mathbf{u}}_p)$, where $\hat{\mathbf{u}}_p$ is normal to $\hat{\mathbf{u}}_q$. The sphere expands uniformly so that the magnitude of r_n remains equal to $2q \cos\alpha$, where the angle α specifies various galaxy locations after t_o when gravitational forces became effective. Expression 1 provides the magnitudes of r_n and v_n, with increases in the sphere radius q producing increases in r_n.

$r_n = 2q\cos\alpha$, where r_n is the separation between the nth galaxy mass M_n and the Milky Way mass M_m, v_n is their separation velocity,
$v_n = 2\,(dq/dt)\cos\alpha$, $q = \int v_q\,dt$ for the radial expansion rate
$= r_n\,(dq/dt)/q$, $v_q = dq/dt$, the declination of M_m to M_n is $\varphi = \pi - \alpha$, and t is greater than t_o. (1)

A Milky Way mirror galaxy offsets the force component $F_n \sin\alpha\, \underline{\hat{\mathbf{u}}}_p$ and doubles $F_n \cos\alpha\, \underline{\hat{\mathbf{u}}}_q$, as indicated in Figure 11.2. By coupling galaxies of mass M_{i+} with counterparts M_{i-}, we see that the surviving forces for the nth galaxy lie along its negative $\underline{\mathbf{k}}_n$ axis as multiples of $\cos\alpha$. The sum over α with q held constant is indicated by Σ_α in equation 2, where $M_s = M_n + M_i = 2\rho\, dA = -4M_t\,(\sin\alpha\cos\alpha)\,d\alpha$, as α varies from $\pi/2$ to 0 for the mass pairs. Integration over α results in the second line of equation 2, where g_o is the gravitational constant. Upon multiplying by $M_n v_q$ and integrating over $dq = v_q\,dt$, we obtain the third line, which includes a constant energy term E_n.

$$dv_q/dt = -2g_o\,\Sigma_\alpha M_s(\cos\alpha)/r_n^2 = 4g_o\int(\cos\alpha)\,\rho\,dA/(2q\cos\alpha)^2,$$
$$= M_t\,g_o\int\sin\alpha\,d\alpha/q^2 = -M_t\,g_o/q^2, \quad M_t \text{ is mass of the universe, and}$$
$$M_n v_q^2/2 - M_n M_t\,g_o/q = E_n, \qquad E_n \text{ is a constant energy,} \qquad (2)$$

$$dq/dt = (2M_t\,g_o)^{1/2}/q^{1/2} = (2/3)\,(9M_t\,g_o/2)^{1/2}/q^{1/2},$$
$$q = (9M_t\,g_o/2)^{1/3}\,t^{2/3}, \qquad \text{when } E_n \text{ is negligible,} \qquad (3)$$

$$r_n = 2q\cos\alpha = (2\cos\alpha)\,(9M_t\,g_o/2)^{1/3}\,t^{2/3},$$
$$v_n = 2\,(dq/dt)\cos\alpha = [(dq/dt)/q]\,r_n = (2/3)\,r_n/t = H_o\,r_n. \qquad (4)$$

If the constant energy term E_n is negligible at t_o, we may re-write equation 2 as 3. Substituting for $(dq/dt)/q$ in expression 1, we obtain Hubble's law as $H_o = (2/3)/t$ in expression 4, to the degree that two given galaxies share the same value of q. The form $H_o = (2/3)/t$ appears to us as a constant since $1/t$ varies so slowly over billions of years. The $\cos\alpha$ factor is critical in specifying galaxies separations, for without it, all galaxies having the same mass (as we have modeled) would be separated from us by the same r_n distance and would recede at the same v_n velocity at any time. But since the

$\cos\alpha$ factors are imbedded in r_n, they do not appear in Hubble's law.

Consistent with other theories, we assume that Newton's law of gravity did not apply initially within the large plasma that followed the big bang. Nevertheless, once it became effective at t_o, continuity requires the values of q_o and the radial velocity v_{qo} of the expanding shell at t_o in the gravity model to match those of the plasma. A NASA source estimates that stars began forming about 400 million years after the big bang, and that this phase lasted 100 to 200 million years. † We will therefore estimate t_o at 600 million years and v_{qo} at the speed of light, $c \cong 3 \times 10^5$ kilometers per second. When we compute v_q from expression 5, we find that setting $M_t \cong 5.75 \times 10^{51}$ kilograms †† provides $v_{qo} \cong c$ at t_o. We further obtain $q_o \cong 277$ MPc $\cong 903$ million light-years at t_o. Our choice for M_t provides Doppler shifts consistent with recession velocities approaching c for distant galaxies whose emissions are just now reaching us, as will be further discussed below.

$$v_q = (2/3)\,(9M_t\, g_o/2)^{1/3}\, t^{-1/3}, \qquad M_t \cong 5.75 \times 10^{51} \text{ kilograms,}$$
$$v_{qo} \cong c, \qquad \text{at } t_o \cong 0.6 \times 10^9 \text{ years.} \qquad (5)$$

Evaluating expression 3 at the present time of $t_p = 13.8$ billion years, we obtain $q_p \cong 2{,}233$ MPc $\cong 7.28$ billion light-years and $v_{qp} = 1.06 \times 10^5$ kilometers per second $\cong 0.352c$. If we select the example of an nth galaxy located at $r_n = 200$ MPc from the Milky Way and use $r_n = 2q_p \cos\alpha$, the $\cos\alpha$ factor would be about 0.0448. Computing $v_n = 2v_{qp} \cos\alpha$, we find that $v_n \cong$ 9,450 kilometers per second, which value lies in the midst of the cluster at r_n in Figure 11.1, as expected. The limited data in Figure 11.1 indicate that $t_p =$ 13.3 billion years, but $t_p = 13.8$ billion years based on matter formation theories with $H_o \cong 47.4$ kilometers per second per mega parsec is equally viable. We add the caveat that v_{qo} might not be the speed of light c at t_o.

---

† See, http://hubblesite.org/hubble_discoveries/science_year_in_review/pdf/2010/the_search_for_the_earliest_galaxies.pdf. The article also reports an observed Doppler shift with $z \cong 10$ for a very distant galaxy.

†† Compare, Davies, P., *The Goldilocks Enigma* (2006), First Mariner The Books, p. 43, estimating the total mass of the observable universe at $M_t = 10^{53}$ kilograms.

Our results show that the 2/3 factor in Hubble's law follows from the time dependent solution of Newton's gravitational attraction law, which supports uniform expansion of a spherical shell of matter into a vacuum, rather than an average universe density in the Friedmann equations. However, our form for r_n is correct only so long as (1) galaxies share the same q-value, (2) non-central forces are not present, and (3) E_n/M_n is negligible.

Collapse Follows Expansion When E_n Is Slightly Negative

Let us regard E_n in equation 2a as a small energy term involving physical processes not modeled. Its form indicates that not all gravitational energy has been converted into v_q as changes in the radius of the universe occur. See Figure 11.2. If, however, E_n/M_n is much smaller than the modeled terms during the evolution process, setting $v_q^2/2 = M_t g_o/q$ will be a good approximation. Taking the derivative of equation 2a and dividing by v_q, we obtain the gravitational force provided by the second line of equation 2. However, the force equation does not specify the behavior of v_q when it becomes zero or undergoes a discontinuity at critical values of q.

$E_n/M_n = v_q^2/2 - M_t g_o/q$, where q is the universe radius at t, v_q is the rate of change of q, M_t is the universe mass, g_o is the gravitational constant, M_n is the nth galaxy mass, and E_n/M_n is a small term. (2a)

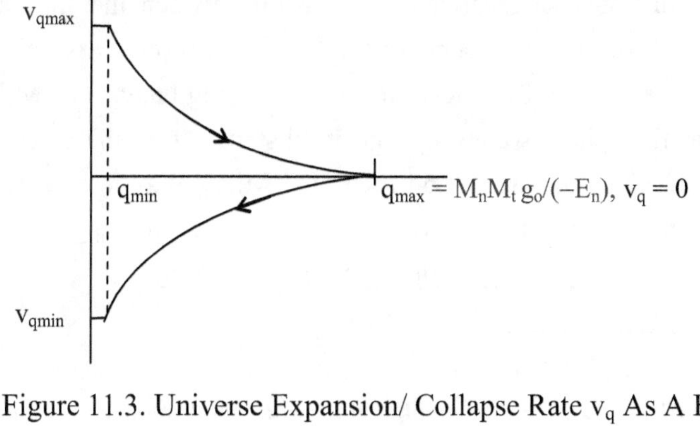

Figure 11.3. Universe Expansion/ Collapse Rate v_q As A Function Of q

If E_n/M_n is a small negative number, $M_t g_o/q$ will become $-E_n/M_n$ at a future maximum of q_{max} with v_q becoming zero and reversing its sign. Gravitational attraction then causes v_q to become more negative as q decreases. When q reaches $q_{min} \cong$ 600 million light-years with $v_{qmin} \cong -c$, matter dissociation produces a super-hot plasma which explodes, and the process starts over with a period of many billions of years. Figure 11.3 depicts v_q as a function of q, where q is increasing and v_q is falling at the present time.

Emissions Of Light Rays At Past Times

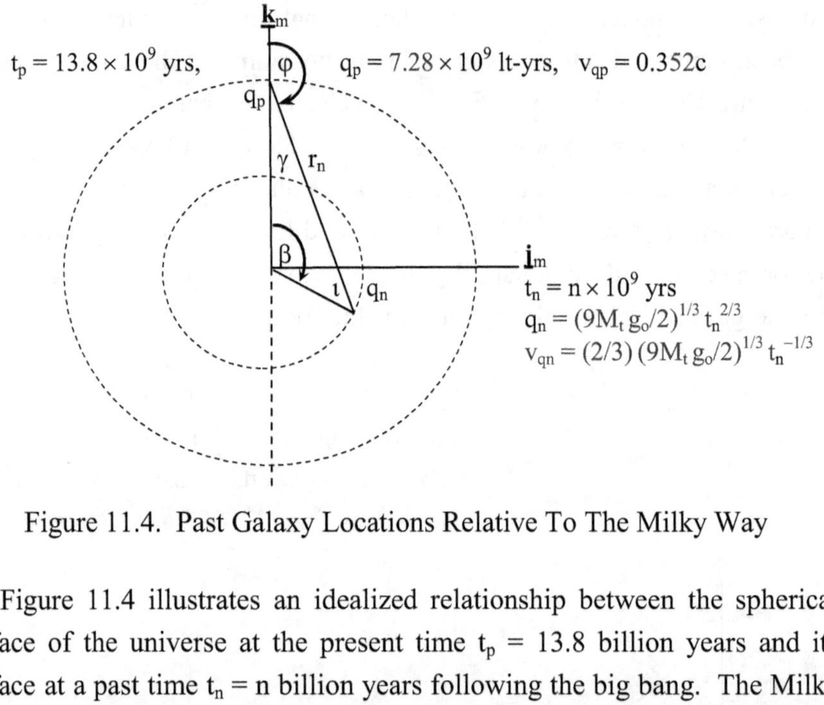

Figure 11.4. Past Galaxy Locations Relative To The Milky Way

Figure 11.4 illustrates an idealized relationship between the spherical surface of the universe at the present time t_p = 13.8 billion years and its surface at a past time t_n = n billion years following the big bang. The Milky Way is located at the sphere's apex $\mathbf{q}_p$, and its distance from the big bang origin is q_p. Expressions 6a through 6d relate the Milky Way to a selected galaxy located on the sphere of radius q_n at t_n. <u>The relationship $r_n = 2q \cos\alpha$ no longer holds since the q-values are different and γ is not α.</u> †

† Relativists agree that Hubble's law does not hold for distances beyond a billion light-years ($\cong$ 307 MPc) and combine Doppler shifts with theory models to estimate r_n. See, *e.g.*, https://starchild.gsfc.nasa.gov/docs/StarChild/questions/redshift.html.

$t_n \cong n \times 10^9 \times 3.16 \times 10^7$ sec, elapsed time in seconds after t_o, (6a)

$q_n \cong 120. \times 10^9\, t_n^{2/3}$ km, n^{th} galaxy distance from origin at t_n, (6b)

$t_p \cong 13.8 \times 3.16 \times 10^{16}$ sec, present time in seconds after t_o, (6c)

$q_p \cong 7.28 \times 10^9$ lt-yrs, Milky Way location on $\underline{\mathbf{k}}_m$ at t_p, (6d)

$r_n^2 = q_p^2 + q_n^2 - 2q_p q_n \cos\beta$, for the angle β between $\underline{\mathbf{q}}_p$ and $\underline{\mathbf{q}}_n$, (6e)

$t_{tvl} = r_n/c = t_p - t_n$, time for light to travel from $\underline{\mathbf{q}}_n$ to $\underline{\mathbf{q}}_p$. (6f)

At t_n the nth galaxy is separated from the origin at a distance q_n, and its radial vector $\underline{\mathbf{q}}_n$ is inclined at an angle β to the $\underline{\mathbf{k}}_m$ axis in expression 6e when it emits a ray of light. After a time lapse of $t_{tvl} = t_p - t_n$ the ray has traveled a distance of r_n, as in expression 6f. In order for the Milky Way to observe this emission at t_p, the value of r_n in 6f must be the same as specified by 6e.

Let us examine the relationships at t_1 = 1 billion years, when universe expansion was driving a galaxy along the $\underline{\mathbf{i}}_m$ axis, *i.e.*, $\beta = \pi/2$. (The entire universe is located on a shell of radius q(t) at any time, where $\pi/2 \le \varphi \le \pi$ and $0 \le \theta \le 2\pi$.) Equation 3 provides q_1 = 1.27 billion light-years at t_1, where r_{n1} = 7.39 billion light-years. Thus, a ray emitted at r_{n1} and traveling at the speed of light c arrived at q_p after 8.39 billion years, and was long gone when the Milky Way arrived at its present location.

The source continued its path along $\underline{\mathbf{i}}_m$, and at $t_2 \cong 5.5$ billion years its parameters were $v_{q2} = 0.478c$ and $q_2 \cong 3.94$ billion light-years, so that γ = arctan(3.94/7.28) $\cong$ 28.4 degrees in Figure 11.4. The angle ι is then 61.6 degrees for β = 90 degrees. The distance traveled by the ray from $\underline{\mathbf{q}}_2$ on $\underline{\mathbf{i}}_m$ to reach $\underline{\mathbf{q}}_p$ is $r_{n2} \cong 8.3$ billion light-years. It follows that a ray emitted at t_2 by the nth galaxy at $\underline{\mathbf{q}}_2$ on the $\underline{\mathbf{i}}_m$ axis arrives at $\underline{\mathbf{q}}_p$ at the same time that the Milky Way arrives. We therefore see the galaxy as it was 8.3 billion years in the past, *i.e.*, 5.5 billion years after the big bang. For a galaxy lying on the negative $\underline{\mathbf{k}}_m$ axis, the requisite value of t_2 for emission of a viewable ray is less, but we still will not be able to see a ray emitted at $\underline{\mathbf{q}}_o$ or at $\underline{\mathbf{q}}_1$.

A ray from a galaxy located on q_n moving with a velocity of v_{nr} along r_n relative to the Milky Way exhibits a Doppler shift of $\lambda_{nr}/\lambda_o = 1/(1 - v_{nr}/c)$, where λ_o is the natural wavelength and λ_{nr} is the wavelength observed at q_p. The separation velocity along r_n is $v_{nr} = v_{qp}\cos\gamma + v_{qn}\cos\iota$, where γ and ι are identified in Figure 11.4, and the velocity components are signed. The result is $v_{nr} = 0.352c \times \cos 28.4° + 0.478c \times \cos 61.6° \cong 0.538c$ for $q_2 \cong 3.94$ billion light-years along $\underline{\mathbf{j}}_m$ at $t_2 \cong 5.5$ billion years. This velocity differs from $v_{nr} = H_o r_n = 0.595c$ for $r_n \cong 8.3$ billion light-years with H_o set to 70 kilometers per second per MPc in an updated form † of Hubble's law. The observed Doppler shift for $v_{nr} = 0.538c$ is $\lambda_{nr}/\lambda_o = 1/(1 - v_{nr}/c) \cong 2.16$. At emission times greater than $t_2 \cong 5.5$ billion years, a ray will timely arrive at $\underline{\mathbf{q}}_p$ if it is emitted in the positive $\underline{\mathbf{k}}_m$ quadrants of Figure 11.3 at some value of γ greater than 28.4 degrees and r_n less than 8.3 billion light-years.

As a further example, let us consider $t_3 = 3.56$ billion years for a galaxy located on the negative $\underline{\mathbf{k}}_m$ axis of Figure 11.3. The value of q_3 at t_3 is about 2.95 billion light-years when a ray is emitted on the negative $\underline{\mathbf{k}}_m$ axis, and it must travel 10.23 billion light-years to reach $q_p = 7.28$ billion light-years. Thus, the total time elapsed is about 13.8 billion years, as required. Since v_{qn} is approximately 0.552c, v_{nr} is about 0.904c. The Doppler shift is $\lambda_{nr}/\lambda_o = 1/(1 - v_{nr}/c) \cong 10.4$, in near agreement with the maximum observed shift.

As discussed above, relativists have misstated the classical Doppler shift as $\lambda_{nr}/\lambda_o = 1 + v_{nr}/c$. The two forms are nearly the same if v_{nr} is significantly less than c, but differ greatly as v_{nr} approaches c, where λ_{nr}/λ_o increases without bound. Relying on the mistake, advocates claim that we are observing recession velocities v_{nr} much greater than c and that space itself is expanding. Both assertions are wrong. We are observing Doppler shifts of $\lambda_{nr}/\lambda_o = 1/(1 - v_{nr}/c)$ for velocities less than c instead of velocities greater than c, and empty space is simply empty space. If future observations reveal values of λ_{nr}/λ_o greater than 10.4, the initial inputs to our model may be changed.

---

† See, The LIGO Scientific Collaboration and The Virgo Collaboration (2017), "A gravitational-wave standard siren measurement of the Hubble constant", *Nature*, advance online publication, doi: 10.1038/nature24471, ISSN 1476 – 4687.

A Viable Theory For Spiral Galaxies

A seminal analysis of the surface brightness of spiral disc galaxies was published by K. C. Freeman in 1970. † Freeman relied on earlier efforts by G. de Vaucouleurs and others which show that the mass densities in the outer galaxy regions, taken to be proportional to brightness, fall off exponentially with distance "r" from the galactic centers. His effort focused on constant scale lengths that vary from about 1 to 6 kiloparsecs (kpc) over the galaxies, and concluded that the source of the densities "remains uncertain".

Previously in 1925, Bertil Lindblad had pointed out that spiral galaxy configurations appear to be unstable and indicate winding problems for the arms. He suggested regions of enhanced density, such as could be provided by waves in gas and dust rotating more slowly than the stars. In 1964 C. C. Lin and F. H. Shu extended Lindblad's proposal †† by modeling a small amplitude wave with a fixed angular velocity revolving in the galaxy at a speed different from the stars. They included correlated elliptical orbits for the stars, moving into and out of the spiral arms. The mechanisms for creating and maintaining such correlations are, however, not specified.

Observing that the velocities of stars in the discs appear to be almost constant over an extensive range of r-values, Vera Rubin ††† in 1983 proposed continuous increases in galaxy mass M_r proportional to the spherical form of r. Since the brightness data indicate an insufficient mass for the Rubin model, a concept of *dark matter* in the spiral discs has been embraced by proponents of that theory. However, the model ignores the observed

---

† See, Freeman, K.C., "On The Disks Of Spiral And S_0 Galaxies" (1970), The Astrophysical Journal, Vol. **160**, p. 811.

†† See, Lin, C.C. and Shu, F.H. (1964), "On The Spiral Structure Of Disk Galaxies", The Astrophysical Journal, Vol. **140**, p. 646. See also, http: //rspa.royalsociety publishing.org/content/465/2111/3425, Anderson, E., and Francis, C. (2009), "Galactic Spiral Structure", Proc.R.Soc.A., **465**, 3425.

††† See, http://www.jstor.org/stable/1691298, Rubin, Vera C., "The Rotation of Spiral Galaxies" (1983), *Science*, Vol. 220, No. 4604, pp. 1339-1344.

exponential mass distribution and does not explain the spiral behavior.

In the alternative, a viable model for spiral disc structures should:

1. Specify average mass densities of $\rho(r) = \rho_o \exp(-r/a)$ for the discs, where r is the distance from the galactic center and ρ_o is a constant;
2. Identify the constant scale length a;
3. Specify the spiral configurations of the galaxies;
4. Provide almost constant orbit velocities over a range of r-values.

Lindblad, Lin, and Shu correctly perceived the influence of waves on the disc structures. But the wave source is the collective gravitational influence of stars and dust interior to an orbit location, rather than the pressure and the assumed magnetic field of interstellar gas and dust.

$$\nabla^2 F = (1/c_g^2)\, \partial^2 F/\partial t^2. \qquad (7)$$

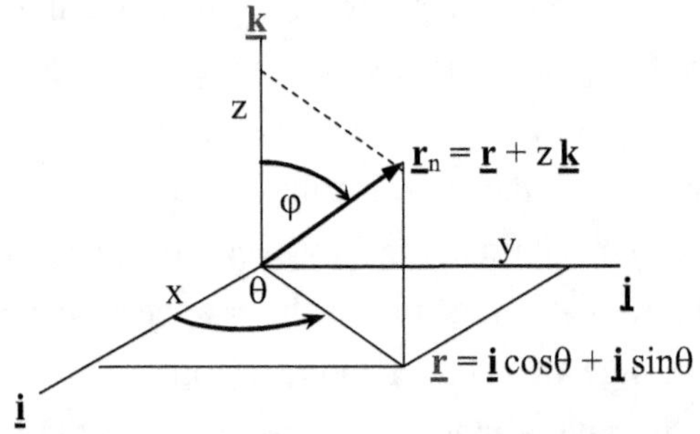

Figure 11.5. Radial Coordinates In A Cartesian Reference Frame

Nature has provided equation 7 as the basic mathematical form for a wave moving at a speed c_g to propagate its influences on the state function $F(\underline{r}_n, t)$ at a time t and location $\underline{r}_n = x\underline{i} + y\underline{j} + z\underline{k}$ in the Cartesian frame depicted in Figure 11.5. The form $\nabla^2 = \nabla \bullet \nabla = \partial^2/\partial x^2 + \partial^2/\partial y^2 + \partial^2/\partial z^2$ is the *Laplacian* operating on F, where $\partial^2 F/\partial x^2$ is the second partial derivative of F with respect to x, etc., the gradient ∇ is $\underline{i}\,\partial/\partial x + \underline{j}\,\partial/\partial y + \underline{k}\,\partial/\partial z$, and "•" indicates the scalar vector product. For simple waves $F(\underline{r}_n, t)$ is the displacement at $\underline{r}_n$, but it may model more involved states in a field such as gravity.

Let us expand equation 7 to model standing gravity wave states in the spiral galaxies. We first multiply equation 7 by $-\sigma_r^2/2$, where σ_r is a specific angular momentum constant at r_n set by the source for a given state. The dimension of $-\sigma_r^2\nabla^2/2$ is specific energy, and is half the square of the transform of velocity $\underline{\mathbf{v}}$ in a Fourier relationship with $\underline{\mathbf{r}}_n$ based on $\exp(i\,\underline{\mathbf{v}} \cdot \underline{\mathbf{r}}_n/\sigma_r)$. In coordinate space v_x appears as $-i\,\sigma_r\,\partial/\partial x$, etc., where $i = (-1)^{1/2}$.

We next include the specific † potential energy, $V_r = -M_r g_o/r_n$, where M_r is the attracting mass interior to $r_n = |\underline{\mathbf{r}}_n| = (x^2 + y^2 + z^2)^{1/2}$ and g_o is the universal gravitational constant. Equation 8 results if we write $F(\underline{\mathbf{r}}_n, t)$ in the form $\Psi(\underline{\mathbf{r}}_n)\,\Im(t)$ and divide by $F(\underline{\mathbf{r}}_n, t)$. Since the two sides of the equation are functions of separated independent variables, each must be equal to a constant we identify as E_r, the specific energy for a state at $\underline{\mathbf{r}}_n$ in the field.

$$[-\sigma_r^2\,\nabla^2\Psi/2 - (M_r g_o/r_n)\Psi]/\Psi = E_r = [-\sigma_r^2\,\partial^2\Im/\partial t^2/(2c_g^2)]/\Im,$$
where σ_r is the specific angular momentum multiplier at r_n, (8)

$$\nabla^2\Psi = (1/r)\,\partial/\partial r\,(r\,\partial\Psi/\partial r) + (1/r^2)\,\partial^2\Psi/\partial\theta^2 + \partial^2\Psi/\partial z^2,$$
where $\underline{\mathbf{r}}_n = r\,(\underline{\mathbf{i}}\cos\theta + \underline{\mathbf{j}}\sin\theta) + z\underline{\mathbf{k}} = r\,\underline{\hat{\mathbf{u}}}_r + z\underline{\mathbf{k}}$,
$\underline{\hat{\mathbf{u}}}_\theta = -\sin\theta\,\underline{\mathbf{i}} + \cos\theta\,\underline{\mathbf{j}}$. (9)

The solution for Ψ is facilitated by expressing the Laplacian in the circular cylindrical coordinates r, θ, and z of expression 9. See Figure 11.5, where $\underline{\mathbf{r}}$ lies in the $\underline{\mathbf{ij}}$ plane and θ is the angle between $\underline{\mathbf{r}}$ and the $\underline{\mathbf{i}}$ axis.

Before attempting to solve equation 8, let us further include the simple vector potential $\mathbf{A}_g$ of expression 10, where $\underline{\hat{\mathbf{u}}}_\theta = -\sin\theta\,\underline{\mathbf{i}} + \cos\theta\,\underline{\mathbf{j}}$ is a unit vector in the θ-direction in the $\underline{\mathbf{ij}}$ plane and s_A is a field constant to be specified. We regard $\underline{\mathbf{A}}_g$ as being produced by the steady-state motions of stars interior to r_n. Since $\underline{\mathbf{A}}_g$ has the dimension of velocity, let us add $i\,\underline{\mathbf{A}}_g$ to the term $i\,\sigma_r\nabla$ before taking the square of the gradient ∇ in equation 11.

$$\underline{\mathbf{A}}_g = s_A\,\sigma_r\,(-\sin\theta\,\underline{\mathbf{i}} + \cos\theta\,\underline{\mathbf{j}})/r_n = s_A\,\sigma_r\,\underline{\hat{\mathbf{u}}}_\theta/r_n, \text{ where } s_A \text{ is a constant,} \quad (10)$$

$$(\sigma_r\nabla + \underline{\mathbf{A}}_g) \cdot (\sigma_r\nabla + \underline{\mathbf{A}}_g)/2 + (M_r g_o/r_n + E_r)\,\Psi = 0, \quad (11)$$

$$[1/(rR)]\,[d/dr\,(r\,dR/dr)] + (1/\Theta)\,[(1/r^2)\,d^2\Theta/d\theta^2 + 2s_A\,(d\Theta/d\theta)/(r_n r)]$$
$$+ s_A^2/r_n^2 + (1/Z)\,d^2Z/dz^2 + (2/\sigma_r^2)\,(M_r g_o/r_n + E_r) = 0, \quad (12)$$

† The term *specific* angular momentum refers to angular momentum divided by the mass of an object in the field, and likewise for *specific* energy.

$d^2Z/dz^2 = 0$, Z is constant for $0 \le z \le h_z$, and Z is zero otherwise. (13)

The divergence $\nabla \cdot \underline{\mathbf{A}}_g$ is zero, and $2\underline{\mathbf{A}}_g \cdot \nabla\Psi$ becomes $2s_A(\partial\Psi/\partial\theta)/(r_n r)$. Writing Ψ as a product of functions of r, θ, and z, *i.e.*, $\Psi = R(r)\,\Theta(\theta)\,Z(z)$, we obtain equation 12 after dividing equation 11 by $R\Theta Z$. Since the term $(1/Z)\,d^2Z/dz^2$ is the only one that depends on Z, † it must be a constant. Taking Z to reflect a bounded mass distribution along $\underline{\mathbf{k}}$, let us set Z to a constant over a disc thickness of h_z that is small in comparison to r, so that $d^2Z/dz^2 = 0$ and $r_n = (r^2 + z^2)^{1/2}$ <u>may be set approximately to r</u> in equation 12.

Multiplying equation 12 by r^2 with d^2Z/dz^2 set to zero, we obtain a collection of terms on the left side of equation 14 that are independent of r and thus determine Θ. When s_A is zero, $(1/\Theta)\,d^2\Theta/d\theta^2 = -m^2$ is a familiar equation with m being an integer or half-integer. If we persist and keep the sum at $-m^2$, equation 14 will be solved by expression 14a, where $i = (-1)^{1/2}$. This is the observed form of mass density in the spiral arms as a function of θ for the disc galaxies. See Figures 11.6 and 11.7 below.

$$(1/\Theta)\,(d^2\Theta/d\theta^2 + 2\,s_A\,d\Theta/d\theta) + s_A{}^2 = -m^2, \tag{14}$$

$$\Theta = \exp(-s_A \pm i\,m)\,(\theta - \theta_o), \text{ where m is an integer or a half-integer, and } \theta_o \text{ is an arbitrary reference angle.} \tag{14a}$$

Figure 11.6 is an artist's conception of the Milky Way structure based on data obtained by NASA's WISE and Spitzer Telescope spacecraft. Figure 11.7 is a view of barred Spiral Galaxy UGC 12158, provided by the Hubble Space Telescope. †† It is located in the constellation Pegasus about 400 million light-years from the Earth. The spiral arms are produced by gravity waves in the galaxy field, wherein the $s_A\theta$ locations represent stable states with stars at other locations being nudged into regions of stability over time.

---

† See, https://www.researchgate.net/publication/1816974_The_radial_scale_length_of_the_Milky_Way (1997), selecting a value of $h_z = 0.2$ kpc and finding that modest variations in h_z did not affect the average density as a function of r.

†† Credit, http://www.spacetelescope.org/images/potw1035a/.

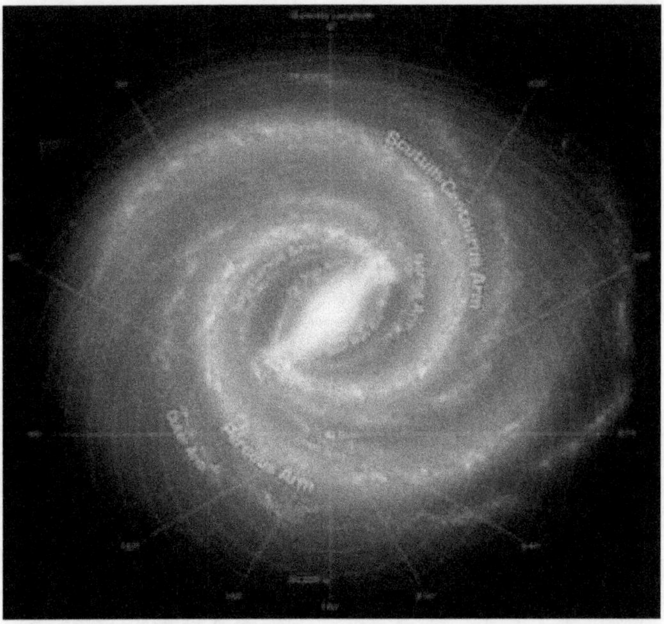

Figure 11.6. Artist's Concept of Milky Way Structure (Credit: NASA)

Figure 11.7. Image of Barred Spiral Galaxy UGC 12158, Taken by NASA/ ESA Hubble Space Telescope's Advanced Camera for Surveys (2010)

We are now left with equation 15 to determine R. Setting $r = ax$ for $a = \sigma_r/(-2E_r)^{1/2}$ and defining a parameter $\chi = M_r g_o a/\sigma_r^2$, equation 15 takes the form of equation 16. Substituting $R(x) = p(x) \exp(-x)$ leads to equation 17, where $R_1(x) = x \exp(-x)$ is a well-behaved solution for $m = \pm 1$ and $\chi = 3/2$.

$$[1/(rR)]\,[d/dr\,(r\,dR/dr)] - m^2/r^2 + (2/\sigma_r^2)\,(M_r g_o/r + E_r) = 0, \tag{15}$$

$$d^2R/dx^2 + (1/x)\,dR/dx + (-m^2/x^2 + 2\chi/x - 1)\,R = 0,$$
$$\text{where } x = r/a,\ a = \sigma_r/(-2E_r)^{1/2}, \text{ and } \chi = M_r g_o a/\sigma_r^2, \tag{16}$$

$$d^2p/dx^2 + (1/x - 2)\,dp/dx + [-m^2/x^2 + (2\chi - 1)/x]\,p = 0,$$
$$\text{for } R(x) = p(x)\exp(-x). \tag{17}$$

Of the many solutions for equation 16, we shall show below that $R_1(x) = x \exp(-x)$ models the observed mass distribution as a function of r in the spiral arms. For $\chi = 3/2$ and $m = 1$, we have $2m^2 M_r g_o a/3 = m^2\sigma_r^2 = r^2 v_\theta^2$, *i.e.*, the square of specific angular momentum for stellar orbits at the varying values of M_r. If the parameter s_A were zero, we would observe a simple decrease in mass density as a function of $R_1(x) = r/a \exp(-r/a)$. But for a given value of $s_A \neq 0$, the stars concentrate about the angular location $s_A\theta$ at a specified value of θ. Separate spirals indicate different values of θ_o and of s_A, which remains unspecified in our theory. Expression 18 provides the composite mass density as a function of r and θ for the observed values of a.

$$\Psi(r, \theta) = (r/a)\exp[-r/a + (-s_A \pm i\,m)\,(\theta - \theta_o)]. \tag{18}$$

The stars remain in nearly circular stable orbits defined by $i\,m\theta$ about the galaxy center. The parameter m may be negative, indicating movement in the clockwise direction in Figure 11.6, as is observed. We attribute the spread about the spiral centroids as being due to interactions among the stars and various orbit eccentricities. Confinement to the $\underline{\mathbf{ij}}$ plane is supported by considering $\underline{\mathbf{B}}_g = \nabla \times \underline{\mathbf{A}}_g$, *i.e.*, the curl of $\underline{\mathbf{A}}_g$. Since $\underline{\mathbf{A}}_g$ consists solely of a $\underline{\hat{\mathbf{u}}}_\theta$ component that yields a constant when multiplied by $r \cong r_n$, its curl will be zero for the two-dimensional configuration. However, small mass currents not confined to the $\underline{\mathbf{ij}}$ plane will produce perturbations in the state solutions.

Using the constant value of a required by the solution for Ψ, we obtain expression 19 for the independent temporal component $\Im(t)$. The solution is a pure oscillation in time based on the constant frequency $\omega_n = c_g(-2E_n)^{1/2}/\sigma_n = c_g/a$, where $-2E_n = \sigma_n^2/a^2$, with $\Im(t)$ repeating after a period of $t_p = 2\pi/\omega_n$.

$$\Im(t) = \sin(\omega_n t + \phi), \qquad \text{where } \omega_n = c_g/a, \text{ and } \phi \text{ is arbitrary.} \qquad (19)$$

The Mass Distribution Of The Milky Way

Our next task is to show that the Milky Way mass distribution increases over an observed range nearly in direct proportion to the cylindrical coordinate r, rather than its spherical counterpart. The mass distribution is based on the galaxy frame coordinates r, θ, and z along **k**, as indicated in Figure 11.8, with unit vectors of $\hat{\mathbf{u}}_r = \cos\theta\,\mathbf{i} + \sin\theta\,\mathbf{j}$, $\hat{\mathbf{u}}_\theta = -\sin\theta\,\mathbf{i} + \cos\theta\,\mathbf{j}$, and **k**.

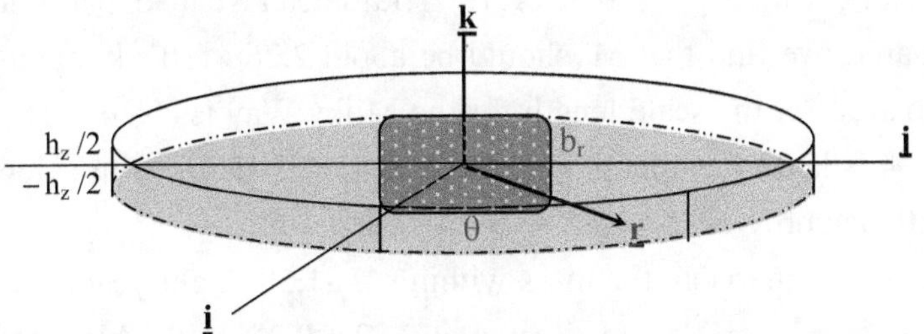

Figure 11.8. Coordinates Used For A Spiral Galaxy Structure

A disk of uniform thickness h_z lies in the **ij** plane beyond a radius of $r = b_r$, with r extending to the maximum galaxy radius. The central bulge mass M_b lies inside b_r, and we will utilize *Newton's shell theorem* to treat gravitational attractions in the galaxy as being due solely to the mass interior to a star's trajectory. The theorem further allows us to model a symmetrical configuration of interior masses as being located at the origin.

For a volume density $\rho(r) = \rho_{\theta z}\exp(-r/a)$ with a scale length a, the mass interior to r is $M_r = \int\!\!\int\!\!\int \rho\, r\, dr\, d\theta\, dz$. If we approximate $\rho_{\theta z}$ as not varying over $-h_z/2 \leq z \leq h_z/2$, and treat Θ as only creating peaks and valleys interior to r without changing the average of ρ as r varies, equation 20 is the averaged integral for M_r evaluated at r. The r-integrand is $R_1(r)$, specified above. It is zero at $r = 0$, maximizes at $r = a$, and returns to zero when $r >> a$.

$$M_r = 2\pi h_z \rho_{\theta z} \int r \exp(-r/a)\, dr,$$
$$= M_m [1 - (1 + x) \exp(-x)], \qquad \text{for } x = r/a. \qquad (20)$$

Computing the integral for $0 \leq x \leq \infty$ identifies $\rho_{\theta z}$ as $M_m/(2\pi a^2 h_z)$, where M_m is the total galaxy mass. This is the same result found by Freeman using Bessel functions. M_r is zero at r = 0, is 0.264 M_m at r = a, and is effectively M_m beyond r $\cong 20a$. The square of specific angular momentum, as provided by $r^2 v_\theta^2 = m^2\sigma_r^2 = 2m^2 M_r g_o a/3$, increases as r increases.

The Constancy Of Stellar Orbit Velocities

The path for the sun has been estimated to lie at an average distance of r $\cong$ 26,100 light-years ($\cong$ 8.01 kpc $\cong 2.47 \times 10^{17}$ kilometers) from the galaxy center with a tangential velocity of $v_\theta \cong 240$ kilometers per second. † Setting $v_\theta^2/r = M_r g_o/r^2$, where $g_o \cong 6.67 \times 10^{-20}$ kilometers cubed per kilogram second-squared, we find that M_r should be about 2.26×10^{41} kilograms. A recent estimate †† of the scale length for the Milky Way is $a = 4.9 \pm 0.4$ kpc. Using a = 4.9 kpc, we obtain x = 8.0/4.9 $\cong$ 1.63 at the sun's location. Equation 20 then provides $M_m \cong 4.66 \times 10^{41}$ kilograms.

By way of comparison, the mass within r = 81,500 light-years, based on Gaia's billion star map, ††† is estimated at $2.32 \pm 0.24 \times 10^{11}\ M_\oplus$, including gas and dust, but without any *dark matter*. Approximating the solar mass at $M_\oplus \cong 1.99 \times 10^{30}$ kilograms leads to $M_m \cong 4.83 \pm 0.24 \times 10^{41}$ kilograms as a data value, which agrees quite well with our theory computation.

---

† See, https://phys.org/news/2012-10-mass-dark-revealed-precise-milky.html. Table I below references data provided by the Very Long Baseline Interferometry (VLBI) of the National Astronomical Observatory of Japan (NAOJ) in the 2012 report.

†† See, Rajkafle, P.; Sharma, S.; Lewis, G.; Bland-Hawthorn, J., "On the Shoulders of Giants: Properties of the Stellar Halo and the Milky Way Mass Distribution" (2014), https://arxiv.org/pdf/1408.1787.pdf, offering $a = 4.9 \pm 0.4$ kpc.

††† See, E. Li (2017), "Modelling mass distribution of the Milky Way galaxy using Gaia's billion-star map", https://arxiv.org/ftp/arxiv/papers/1612/1612.07781.pdf.

Multiplying expression 20 by g_o/r, we obtain equation 21 to specify v_θ^2 for a circular orbit as a function of x, which maximizes at r ≅ 1.8*a*. Using *a* = 4.9 kpc and $M_m \cong 4.66 \times 10^{41}$ kilograms, we find that v_θ reaches its maximum of about 248 kilometers per second at r ≅ 28,700 light-years.

$$v_\theta^2 = (g_o M_m/a)\,[1 - (1 + x)\exp(-x)]/x. \qquad (21)$$

Table 8. Star Orbit Velocities Vs. Distances To The Milky Way Center

Distance r to MW center in light-years	Mass M_r interior to r in 10^{41} kg	Theory: v_θ at r in km/sec	Avail. data: v_θ at r in km/sec *
6,500 (2 kpc)	0.30	125	no data
10,000	0.61	207	190
12,500	0.86	220	185 – 210
14,000	1.02	226	210
20,000	1.66	241	190 – 245
23,000	1.97	245	240 – 250
26,100	2.29	247	240 ± 14
28,700	2.50	248	210 – 240
30,000	2.61	247	205 – 250
34,000	2.93	246	240
38,000	3.19	243	no data
43,000	3.49	229	250
50,000	3.81	224	no data
100,000	4.61	175	no data

* A data cluster occurs for $225 \le v_\theta \le 250$ kilometers per second at $18{,}500 \le r \le 35{,}000$ light-years.

Relying on equation 21, and our model values for M_m and *a*, we obtain the results for v_θ displayed by Table 8. The calculations begin at r = 2 kpc, which is the bulge radius used for the Gaia data. The table summarily shows that v_θ remains between 220 and 248 kilometers per second as r varies from 12,500 to 50,000 light-years. The last two columns compare our theory results with available data. Different observed velocities at the same value of r reflect a mixture of data collection/ reduction issues, eccentric orbits, and angular dependencies. Including more dust and gas would result in higher values of v_θ, such as 270 kilometers per second which has been estimated by the National Radio Astronomy Observatory (NRAO).

Equation 21 has a single maximum at $r_o \cong 28{,}700$ light-years, *i.e.*, $x_o \cong 1.8$. Expanding the equation in a Taylor series about x_o leads to equation 22, where $d^2f(x_o)/dx^2$ is negative. The result for v_θ thus represents a modest downward opening parabola about x_o at this value. Figure 11.9 is a rough sketch of its behavior. Its rise beginning at $r \cong 6{,}500$ light-years is observed, but there exists a gap in the data from $r \cong 34{,}000$ to $43{,}000$ light-years. No data are available beyond the latter value for a galaxy whose diameter is at least 4 times this value. In any case, there is no *dark matter* mystery here. †

$$f(x) = v_\theta^2 a/(g_o M_m) \cong f(x_o) + [d^2f(x_o)/dx^2]\,(x - x_o)/2. \qquad (22)$$

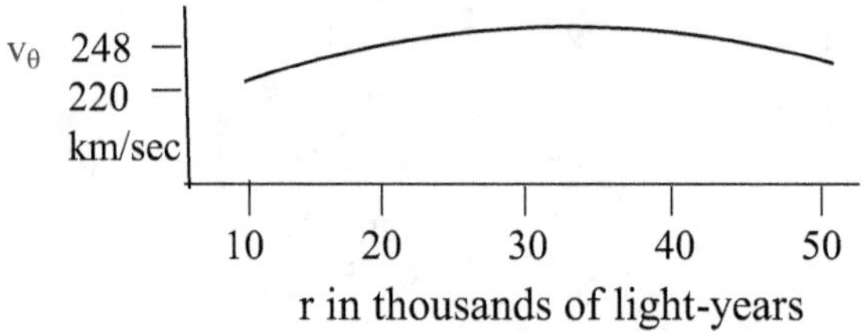

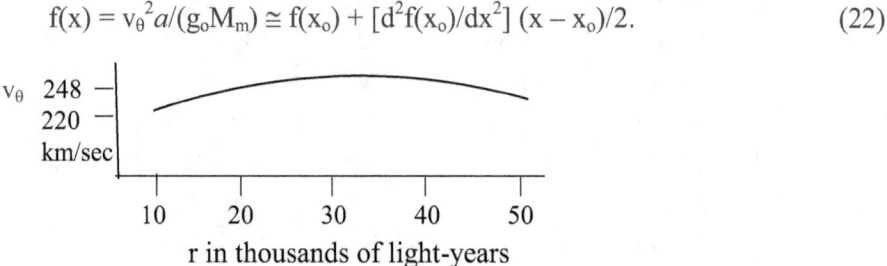

Figure 11.9. Orbit Velocity v_θ Vs. Distance r To The Milky Way Center

Magnetic-Like Gravitational Forces Mislabeled As Dark Energy

An issue remains concerning the process whereby rotational states have come about from a radially-expanding sphere of matter in the first place. We also need to explain the wide scatter called *peculiar* velocities in the straight-line plot for H_o in Figure 11.1. This same issue appeared in Hubble's 1929 data for individual stars and galaxies at distances within 2 mega-parsecs from the Earth. See Figure 11.10.

---

† See also, McGaugh, S. and Lelli, F., "Radial Acceleration Rotation in Rotationally Supported Galaxies" (2016), Phys.Rev.Ltrs. **117**, 201101, finding for 153 galaxies that "the dark matter contribution is fully specified by that of the baryons". See further, Kosowsky, A., "Connecting the Bright and Dark Sides of Galaxies" (2016), Am. Phys.Soc. **9**, 130, reaching the same conclusion.

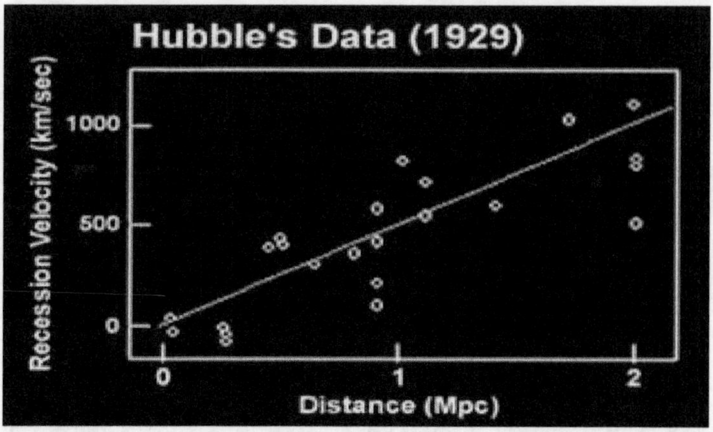

Figure 11.10. Hubble's 1929 Data (Credit PNAS) †

The graph shows several objects near the origin with negative recession velocities that remain negative when the sun's motion has been corrected. The Andromeda galaxy in particular, located about 2.5 million-light years (0.77 MPc) from the Earth, appears to be on a collision course with the Milky Way and is projected to pass through it in about 4 billion years. †† How are these phenomena explained for a continually expanding universe?

Adding to the complexity, relativists argue for modeling Hubble observations proportional to 1/2 by a scale parameter *a*(t) in their equations. †††

> "The Universe has gone through three distinct eras: radiation-dominated …; matter-dominated …; and dark-energy-dominated … The evolution of the scale factor is controlled by the dominant energy form: $a(t) \sim t^{2/3(1+w)}$ (for constant w). During the radiation-dominated era $a(t) \sim t^{1/2}$; during the matter-dominated era, $a(t) \sim$

† See, www.pnas.org/content/101/1/8.full. Hubble's data miscalculate the sun's motion, and recent analyses have lowered the slope of his line significantly.

†† See, http://www.nasa.gov/mission_pages/hubble/science/milky-way-collide.html.

††† See, Frieman, J., Turner, M., and Huterer, D. (2008), "Dark Energy and the Accelerating Universe", *Annual Review of Astronomy and Astrophysics*, **46** (1): pp. 385, 426. Also, Friedmann, A., (Tr. 1999), "On the Possibility of a World with Constant Negative Curvature of Space", *General Relativity and Gravitation.* **31** (12): 2001.

$t^{2/3}$; and for the dark energy-dominated era, assuming w = −1, asymptotically $a(t) \sim \exp(Ht)$. … This implies that the Universe began accelerating at $z \cong 0.4$ and age $t \cong 10$ Gyr."

But there is a dispute as to whether the universe expansion rate is accelerating. † Rather than interpreting unsettled issues as evidence of undefined *dark energy* in the Friedmann model, we maintain that negative recessions, peculiar velocities, and a Hubble factor of 1/2 reflect the effects of galaxy-level vector potentials similar to the one we are modeling for the Milky Way.

When the sun had formed some 5 billion years ago, stellar orbits in the Milky Way were creating a galaxy-level $\underline{\mathbf{B}}_{gm}$ flux expressible as $\nabla \times \underline{\mathbf{A}}_{gm}$. The flux produces a non-central force of $M_n\, \underline{\mathbf{v}}_n \times \underline{\mathbf{B}}_{gm}$ on the nth galaxy mass M_n, where $\underline{\mathbf{v}}_n$ is its velocity relative to the Milky Way mass M_m. An equal and opposite force acts on M_m, and many other galaxies produce $\underline{\mathbf{B}}_{gn}$ fluxes.

For galaxies at various locations in a selected Milky Way $\underline{\mathbf{ijk}}$-frame, let us model $\underline{\mathbf{A}}_{gm}$ in the spherical polar coordinates depicted in Figure 11.11 below, where $\underline{\mathbf{r}}_n = r_n\, \underline{\hat{\mathbf{u}}}_{nr}$ and the unit vectors are defined by expression 23.

$$\underline{\hat{\mathbf{u}}}_{nr} = \underline{\mathbf{i}}\cos\theta\sin\varphi + \underline{\mathbf{j}}\sin\theta\sin\varphi + \underline{\mathbf{k}}\cos\varphi,$$
$$\underline{\hat{\mathbf{u}}}_{\varphi} = \underline{\mathbf{i}}\cos\theta\cos\varphi + \underline{\mathbf{j}}\sin\theta\cos\varphi - \underline{\mathbf{k}}\sin\varphi,$$
$$\underline{\hat{\mathbf{u}}}_{\theta} = -\underline{\mathbf{i}}\sin\theta + \underline{\mathbf{j}}\cos\theta, \qquad (23)$$

$$\underline{\mathbf{A}}_{gm} = s_A\, \sigma_m (\sin\varphi)(-\sin\theta\, \underline{\mathbf{i}} + \cos\theta\, \underline{\mathbf{j}})/r_n = s_A\, \sigma_m (\sin\varphi)\, \underline{\hat{\mathbf{u}}}_{\theta}/r_n,$$

where $\underline{\mathbf{r}}_n = r_n\, \underline{\hat{\mathbf{u}}}_{nr}$ is separation between the Milky Way and the nth galaxy, s_A is a field constant, and $\sigma_m = (M_m g_o a_m)^{1/2}$ is a galaxy-level parameter. (24)

The vector potential $\underline{\mathbf{A}}_{gm}$ that we modeled in cylindrical coordinates appears in spherical polar coordinates as expression 24 when r_n is not constrained the lie in the $\underline{\mathbf{ij}}$-plane. The coordinates used for orbits in the Milky Way provide the limiting form of $\underline{\mathbf{A}}_{gm}$ when φ is set to $\pi/2$. Upon computing

† See, https://www.sciencedaily.com/releases/2016/10/161021123238.htm. See also, Overbye, Dennis. "Cosmos Controversy: The Universe Is Expanding, But How Fast?" *New York Times*, 20 February 2017.

$\nabla \times \underline{\mathbf{A}}_{gm}$, we obtain expression 25 † to specify $\underline{\mathbf{B}}_{gm}$ at the nth galaxy location. Without loss of generality, we may select the initial location of the nth galaxy at $\theta = \pi/2$, as in Figure 11.11 below, where θ may subsequently vary. The flux produces a force of $\underline{\mathbf{F}}_B = M_n \underline{\mathbf{v}}_q \times \underline{\mathbf{B}}_{gm}$ on the nth galaxy, where $\underline{\mathbf{v}}_q$ is the expansion velocity vector. Since the dimension of $\underline{\mathbf{B}}_{gm}$ is inverse seconds, $\underline{\mathbf{F}}_B$ is an expression for continual rotation of $\underline{\mathbf{v}}_q$ with a frequency of $\underline{\mathbf{B}}_{gm}$. The force does not affect the magnitude of $\underline{\mathbf{v}}_q$, but only changes its direction.

$$\underline{\mathbf{B}}_{gm} = 2s_A \sigma_m (\cos\varphi)\, \underline{\hat{\mathbf{u}}}_{nr}/r_n^2,$$
$$= 2s_A \sigma_m (\cos\varphi)(\sin\varphi\, \underline{\mathbf{i}} + \cos\varphi\, \underline{\mathbf{k}})/r_n^2, \text{ at } \theta = \pi/2. \qquad (25)$$

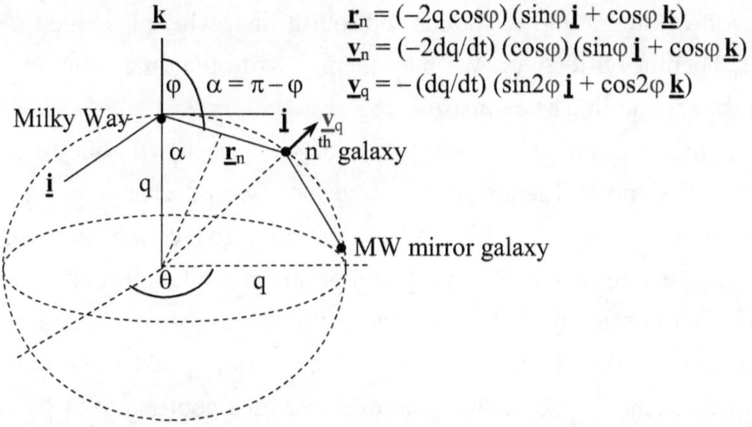

Figure 11.11. Position And Velocity Of The nth Galaxy

With the nth galaxy initially lying in the $\underline{\mathbf{jk}}$-plane of the Milky Way, our model depicted in Figure 11.11 provides $\underline{\mathbf{v}}_q = (-dq/dt)(\sin 2\varphi\, \underline{\mathbf{j}} + \cos 2\varphi\, \underline{\mathbf{k}})$ for the nth galaxy's motion due to universe expansion. Since both $\underline{\mathbf{v}}_q$ and $\underline{\mathbf{B}}_{gm}$ lie in the $\underline{\mathbf{jk}}$-plane, $\underline{\mathbf{F}}_B = M_n \underline{\mathbf{v}}_q \times \underline{\mathbf{B}}_{gm}$ is a force along the $\underline{\mathbf{i}}$-axis, expression 26. Let us use the notation $dv_q/dt|_b$ in equations 27 and 28 to indicate "the magnitude of the directional change in dv_q/dt due to $\underline{\mathbf{B}}_{gm}$," etc.

$$\underline{\mathbf{F}}_B/M_n = \underline{\mathbf{v}}_q \times \underline{\mathbf{B}}_{gm} = -2(dq/dt)(\cos\varphi)(s_A \sigma_m/r_n^2)(\sin\varphi)\, \underline{\mathbf{i}} = d\underline{\mathbf{v}}_q/dt. \qquad (26)$$

---

† See, *e.g.*, Hildebrand, F.B. (1960). *Advanced Calculus For Engineers*, Prentice-Hall, N.J., at p. 329, for computation of the curl in spherical polar coordinates.

$$dv_q/dt|_b = s_A \sigma_m (\sin\varphi) [(dr_n/dt)/r_n^2]|_b,$$
$$v_q|_b = dq/dt|_b = -s_A (\sigma_m/r_n)|_b \sin\varphi,$$
$$-2(\cos\varphi)(dq/dt)|_b = v_n|_b = (dr_n/dt)|_b = s_A (\sigma_m/r_n)|_b \sin 2\varphi,$$
$$r_n^2|_b = (2 s_A \sigma_m \sin 2\varphi)(t - t_b), \qquad \underline{\mathbf{B}}_{gm} \text{ becomes effective at } t_b, \tag{27}$$

$$r_n|_b = (2 s_A \sigma_m \sin 2\varphi)^{1/2} (t - t_b)^{1/2},$$
$$v_n|_b = (1/2)\, r_n|_b/(t - t_b) = H_b\, r_n|_b, \qquad \text{for } t > t_b. \tag{28}$$

When we integrate the first line of expression 27 over time with φ remaining constant, we obtain the second line. Multiplying $v_q|_b$ by the constant $-2\cos\varphi$, we obtain the third line, where $-2(\cos\varphi)(dq/dt)|_b = v_n|_b = dr_n/dt|_b$. Integrating, we obtain the fourth line, where t_b is the time at which $\underline{\mathbf{B}}_{gm}$ becomes effective. We may then re-write the time solution in the form of the second line of expression 28. †

Thus, we have two components of $\underline{\mathbf{v}}_n$. The first is due to universe expansion $\underline{\mathbf{v}}_q$ normal to its spherical surface, whose changing magnitude as a function of time is specified by expression 4 above. The second, expressed as $\underline{\mathbf{v}}_n|_b$, remains orthogonal to $\underline{\mathbf{v}}_q$ and produces continual rotation about the $\underline{\mathbf{k}}$-axis that defines $\underline{\mathbf{B}}_{gm}$. The rotation continues to displace the galaxy from its original angular position, and that is the meaning of the increase with time expressed as $r_n|_b$. A Hubble relationship with a coefficient of 1/2 results, and its observation supports the presence of $\underline{\mathbf{F}}_B$ forces acting on spiral galaxies.

Since $\underline{\mathbf{v}}_n \bullet \underline{\mathbf{v}}_n \times \underline{\mathbf{B}}_{gm} = 0$, $\underline{\mathbf{F}}_B$ cannot increase the kinetic enengy of the galaxies. So long as $2 s_A \sigma_m (dq/dt)$ is much smaller than $M_s g_o$, the effects of $\underline{\mathbf{F}}_B$ will appear as perturbations to the Newtonian force solution and will produce scatter in the Hubble diagrams. †† The end effect on any given galaxy will be determined by the sum of of such forces, which is total-geometry dependent. Some geometrical configurations may act to increase galaxy separations and others may reduce them. For Andromeda and the Milky Way the latter occurs, and the effect is sufficiently strong to indicate a

---

† Since σ_m has the dimension of kilometers squared per second, $\sigma_m^{1/2} t^{1/2}$ is a length.

†† Wave propagation delays for galaxy locations at long past times also cause peculiar velocities, as shown above under *Emissions Of Light Rays At Past Times*.

galaxy collision. $\underline{\mathbf{F}}_B$ should also be expected to play a major role in forming galaxy clusters such as exist for the Local Group containing the Milky Way.

Our theory allows for the onset of the effects of $\underline{\mathbf{B}}_{gm}$ when newer stars such as the sun formed in the arms of spiral galaxies. The creation of this force component several billion years ago has led some astronomers to conclude that the universe expansion rate mysteriously began to increase at that time. But we regard the observed change as evidence of the initiation of $\underline{\mathbf{B}}_{gm}$ forces, rather than the emergence of some form of *dark energy*.

Summarizing this discussion, our simple cosmology model (see Figures 11.2, 11.3, 11.4, and 11.11) has enabled the derivation of Hubble's law from basic physics principles and opposes claims that the rate of expansion of the universe is increasing despite gravitational attraction. Our theory has also shown that *dark matter* is not needed to explain the Milky Way structure. Included in our efforts are a model of the Milky Way's mass distribution consistent with luminosity observations, and a wave equation whose solution defines its spiral geometry as part of a negative exponential state with a constant scale length, as is observed for these galaxies.

We have additionally provided a qualitative explanation for peculiar velocities in the Hubble diagrams, and for the fact that the velocities of some stars and galaxies are directed toward the Milky Way, which could not be so for a constantly expanding universe. We have shown how linear velocities, including those imparted by a big bang, were converted to rotational motion without relying on non gravitational processes. Our efforts constitute a first step toward understanding the phenomenon now labeled as *dark energy*.

Unhappily, the acceptance of our model is unlikely by cosmologists who have invested careers in advocating *dark matter* and *dark energy* proposals as elements in the Friedmann equations of general relativity. We are encouraged, however, by recent publications challenging *dark matter* theories. [†]

[†] See, E. Li (2017), "Modelling mass distribution of the Milky Way galaxy using Gaia's billion-star map", *supra*. See also, McGaugh, S. and Lelli, F. (2016), "Radial Acceleration Rotation in Rotationally Supported Galaxies", *supra*. See further, Kosowsky, A., "Connecting the Bright and Dark Sides of Galaxies" (2016), *supra*.

12 – A SUMMATION OF THE EFFORT

It would be difficult to find any other single concept that would serve to explain so many previously unexplained phenomena as our proposal of $\underline{\mathbf{B}}_g$, a magnetic-like component for the gravitational field. Applying an analogous wave equation of atomic physics to the solar gravitational field, we used the solar mass $M_\oplus$ and a fundamental parameter a_o to define $\sigma = (M_\oplus g_o a_o)^{1/2}$ as the equivalent of the Planck constant $\hbar$ divided by the electron mass m_e. Standing wave solutions that include a vector potential $\underline{\mathbf{A}}_g$ in the solar field allowed us to specify observed orbital parameters for the planets, including their mean radii, inclinations, and spin states. Use of the resulting coordinate frame for the Earth further enabled us to model its spin axis nutation and equinox precession. Our results additionally show the mean Chandler Wobble period for the Earth to be 420 days and predict a long-term period of about 105,000 years, consistent with the Milanković theory for the Ice Ages.

Application of our theory also explains observed anomalies in the trajectories of Pioneer 10 and 11 spacecraft. Unlike the heat radiation model accepted for these anomalies, our theory may be applied to other spacecraft which have exhibited similar unusual behavior, but do not contain comparable heat sources. Using forms of Maxwell's equations for gravitational waves and accepting the speed of light c as their constant velocity, we were led to define gravitational permeability of $\kappa_g = 4\pi g_o/c^2$, where g_o is the universal gravitational constant. We then used κ_g multiplied by the sun's mass $M_\oplus$ to model the perihelion advance for Mercury's orbit as the gravitational equivalent of Larmor precession. This is the same form used by Albert Einstein to derive the Schwarzschild radius $2\kappa_g M_\oplus$ of his general theory.

Our theory's abilities to explain observed planetary orbit radii and their orbit and spin inclinations, as well as the Pioneer anomalies, make it unlikely that our findings regarding the existence of a gravitational vector potential $\underline{\mathbf{A}}_g$ for the solar field could be coincidences. A number scheme, such as the Titius-Bode law, might mimic one data set but not multiple independent ones. We do not claim that our analyses are the only acceptable approaches for these phenomena, but will assert that our results are correct and support reconsideration of the special theory of relativity in light of observations.

As a part of the effort, we have shown that principal experiments claimed to validate the space-time-velocity relationships of the special theory of relativity for constant linear motion do no such thing. Two of the supposed theory validations – the Michelson-Morley experiments and the non-parallax aberration of starlight – are instead explained by modeling an observer's movement during the time that a light ray propagates at the universal speed c. We pointed out that the Michelson-Morley and related experiments support only the constancy of phase for spherical wave fronts, and not the relationship between space and time advocated by the special theory. The components of phase were specifically identified and shown to ensure its constancy when time lapses for observations are multiplied by the velocity of a moving observer. Our results agree with observations.

We concluded by offering a non-exotic cosmology theory, addressing the relationship between galaxy velocities and their distances from the Milky Way. The 2/3 coefficient in the Hubble equation was shown to be the result of modeling galaxy separations of the form $r_n = (9M_sg_o/2)^{1/3}\, t^{2/3}$, where t is the elapsed time following a big bang and M_s is the sum of the Milky Way mass M_m and the nth galaxy mass M_n. For a zero sum energy, the second derivative of r_n evidences the applicability of Newton's gravitational force by itself to universe expansion up to perhaps 8.8 billion years following the big bang. Light rays from distant galaxies that are just now reaching us were emitted at a time when the expansion rate was greater than it is now, and the rate continues to decrease under the influence of gravity.

Turning to the Milky Way structure, we used cylindrical coordinates to model the observed exponential mass distribution and refuted the claim that it is dominated by supposed *dark matter*. Our wave theory for its momentum and energy states reveals the basis for its constant scale length and exponential mass distribution and specifies its spiral arms. Lastly, we proposed a galactic-level magnetic-like flux $\underline{\mathbf{B}}_{gm}$ and showed that its effects based on $\sigma_m = (M_mg_oa_m)^{1/2}$, where a_m is a galactic length parameter, may be used to explain the peculiar velocities in the Hubble graphs. The magnetic-like forces allow some galaxies to repel and others to attract, an option not available under the general theory of relativity. We ended by showing that an attendant $\underline{\mathbf{v}}_n \times \underline{\mathbf{B}}_{gm}$ force results in constant rotation of the changing universe expansion vector and provides a term proportional to 1/2 in the Hubble relationship, which phenomenon has been misinterpreted by relativists.

Our magnetic-like model also provides for formation of stellar structures by other than thermal processes and nominal gravitation attraction among the galaxies. Configurations such as the *Great Wall*, 700 million light years across, have had not had ample time to form under these influences alone. Faced with a *horizon problem* of achieving equilibrium within 13.8 billion years for an isotropic universe, general theory advocates have touted an *inflation* thought experiment, which violates known laws of physics. †

Our approach allows for the possibility of galaxy recession velocities v_n greater than the speed of light c; however, we have concluded that we cannot observe them because their light emissions cannot reach us if v_n exceeds c. Rays with a redshift of z = 3.395 were observed in 1988, and another was later observed for z = 6.3. More recently, a value of z = 10 was detected by the Hubble Telescope. The correct Doppler formula specifies these recession velocities at 0.772c, 0.863c, and 0.909c, respectively, rather than values that exceed the speed of light.

Our wave theory may also be used to support efforts to understand the evolution of the solar system, showing that once the sun's body reached a steady-state configuration, regions of stability existed in its gravitational field for matter to coalesce and form planets. While the theory cannot predict which states in a star's field will be populated, it sets the allowed orbit radii and inclinations based on the properties of the field source.

Throughout the above-referenced analyses, we have questioned the role played by the general theory of relativity as the ultimate theory designed to explain gravitational phenomena. What is its role in explaining the Titius-Bode relationship and the inclinations of the planetary orbits? The Earth's spin obliquity, orbit precession, and the Chandler Wobble? The anomalies in the trajectories of Pioneer and other spacecraft? The enormous Doppler shifts of rays from distant galaxies being observed by the Hubble Telescope, the Spitzer Space Telescope, and WISE? The exponential mass distribution

† See, Marmet, P, "The Cosmological Constant and the Redshift of Quasars", 11/26/2010, http://www.newtonphysics.on.ca/quasars/index.html, for further discussion.

and spiral arms of the Milky Way? Sadly, the general theory remains silent on these issues, except for proposing unrealistic clams of the expansion of space itself, two forms of velocity, dark matter, dark energy, [†] and so forth.

The analyses presented herein are based on proven principles of physics that were in existence before relativity theories were proposed. If portions of our concepts prove to be incorrect, they can be modified, but hopefully not as often as the relativity models, which seem to require major revisions every time a new discovery is made. We note that in 1952 relativity advocates with some authority estimated the age of the universe at about 4 billion years, [††] but its age has since grown to almost 14 billion years. That's quite a feat, and illustrates the necessity for scientists to be ready to modify their theories as new discoveries are made.

On a closing note, we encourage physics students to go online and read about the enduring contributions to our understanding of the laws of physics by great minds who preceded us. Among those which should be included, in addition to Isaac Newton and Albert Einstein, are Leonhard Euler (1707-1783), Joseph-Louis Lagrange (1736-1813), Pierre-Simon Laplace (1749-1827), Carl Friedrich Gauss (1777-1855), and especially Michael Faraday (1791-1867) and James Clerk Maxwell (1831-1879). Our listings of the names and references in the above text to some of their contributions are cursory at best.

[†] Vector potentials may be compared to some degree to Einstein's original proposal of a cosmological constant Λ. He later called Λ his "greatest error", but he was on the right track in the pursuit of a second form of gravitational force.

[††] See, *e.g.*, Gamow, G., *The Creation of the Universe*, A Mentor Book, NY (1952).

APPENDIX 1 [†] – FRAMES AND OPERATORS

Coordinate Systems For Orbits

We could use the Cartesian coordinates x, y, and z of an inertial $\underline{\mathbf{i}}_o\underline{\mathbf{j}}_o\underline{\mathbf{k}}_o$ frame centered in the sun to describe the planetary orbits. [††] However, for forces which are functions of $\underline{\mathbf{r}} = x\,\underline{\mathbf{i}}_o + y\,\underline{\mathbf{j}}_o + z\,\underline{\mathbf{k}}_o$, it is more convenient to use the spherical polar coordinates depicted in Figure A1.1, a duplicate of Figure 2.2 above. The vector $\underline{\mathbf{r}}$ can be expressed as $\underline{\mathbf{r}} = r\,\underline{\hat{\mathbf{u}}}_r$, where $\underline{\hat{\mathbf{u}}}_r$ is the unit vector $\underline{\hat{\mathbf{u}}}_r = \sin\varphi\cos\theta\,\underline{\mathbf{i}}_o + \sin\varphi\sin\theta\,\underline{\mathbf{j}}_o + \cos\varphi\,\underline{\mathbf{k}}_o$, and r is the absolute value of $\underline{\mathbf{r}}$, written as $|\underline{\mathbf{r}}| = (x^2 + y^2 + z^2)^{1/2}$. The reference angles depicted in the figure are φ = arc-cosine(z/r) and θ = arc-tangent(y/x), and we refer to a given location by the values of r, φ, and θ at that location. The unit vectors for φ and θ are $\underline{\hat{\mathbf{u}}}_\varphi = \cos\varphi\cos\theta\,\underline{\mathbf{i}}_o + \cos\varphi\sin\theta\,\underline{\mathbf{j}}_o - \sin\varphi\,\underline{\mathbf{k}}_o$, and $\underline{\hat{\mathbf{u}}}_\theta = -\sin\theta\,\underline{\mathbf{i}}_o + \cos\theta\,\underline{\mathbf{j}}_o$ in terms of the $\underline{\mathbf{i}}_o\underline{\mathbf{j}}_o\underline{\mathbf{k}}_o$ coordinates. The unit vectors have absolute values of 1, and they point in the increasing directions of the variables. Vectors other than $\underline{\mathbf{r}}$ usually require the use of $\underline{\hat{\mathbf{u}}}_\varphi$ and $\underline{\hat{\mathbf{u}}}_\theta$ as well as $\underline{\hat{\mathbf{u}}}_r$. Planet velocity $\underline{\mathbf{v}} = d\underline{\mathbf{r}}/dt$, for example, appears as $\underline{\mathbf{v}} = \underline{\hat{\mathbf{u}}}_r\,dr/dt + \underline{\hat{\mathbf{u}}}_\varphi\,r\,d\varphi/dt + \underline{\hat{\mathbf{u}}}_\theta\,r\,(\sin\varphi)\,d\theta/dt$. The unit vectors do not define an inertial reference frame since they do not remain constant, but our results can be specified in inertial space by substituting for $\underline{\hat{\mathbf{u}}}_r$, $\underline{\hat{\mathbf{u}}}_\varphi$, and $\underline{\hat{\mathbf{u}}}_\theta$ in terms of x, y, and z in an inertial $\underline{\mathbf{i}}_o\underline{\mathbf{j}}_o\underline{\mathbf{k}}_o$ frame.

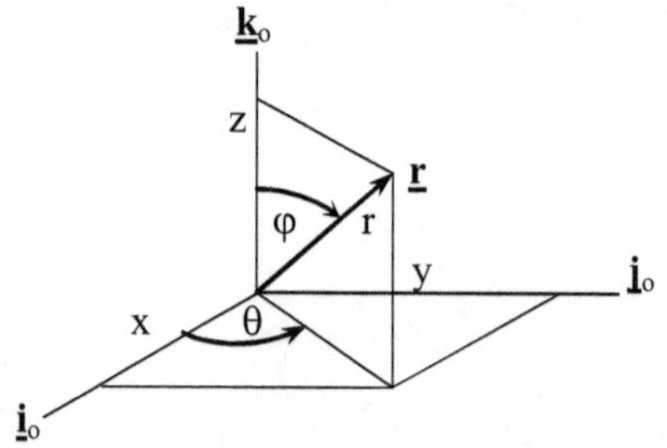

Figure A1.1. Spherical Polar Coordinate System

[†] The purpose of the appendices is to provide mathematical detail in support of the body of the document, and they generally correspond to chapters of the same number.

[††] Albert Einstein used Cartesian coordinates to formulate his special theory.

Vector Operations Involving The Gradient

When solving physics problems we often apply *operators* to vectors and functions. The operator symbols are a shorthand means of expressing processes which have some physical meaning, and they appear in numerous applications. One such form is the *gradient* of a scalar variable V(x, y, z), which is called *del* V and is defined as $\nabla V = \underline{\mathbf{i}}_o\, \partial V/\partial x + \underline{\mathbf{j}}_o\, \partial V/\partial y + \underline{\mathbf{k}}_o\, \partial V/\partial z$. Another is the inner (or scalar) product of the gradient operator ∇ and any vector $\underline{\mathbf{F}} = F_x\, \underline{\mathbf{i}}_o + F_y\, \underline{\mathbf{j}}_o + F_z\, \underline{\mathbf{k}}_o$, written as $\nabla \bullet \underline{\mathbf{F}} = \partial F_x/\partial x + \partial F_y/\partial y + \partial F_z/\partial z$ and called the *divergence* of $\underline{\mathbf{F}}$, or *del dot* $\underline{\mathbf{F}}$. When $\underline{\mathbf{F}}$ is the gradient of a scalar V, the form of $\nabla \bullet \nabla V$ is $\nabla^2 V = \partial^2 V/\partial x^2 + \partial^2 V/\partial y^2 + \partial^2 V/\partial z^2$, or *del-squared* V, and ∇^2 is called the *Laplacian operator*. The curl of a vector $\underline{\mathbf{A}}$, is $\nabla \times \underline{\mathbf{A}} = (\partial A_z/\partial y - \partial A_y/\partial z)\, \underline{\mathbf{i}}_o + (\partial A_x/\partial z - \partial A_z/\partial x)\, \underline{\mathbf{j}}_o + (\partial A_y/\partial x - \partial A_x/\partial y)\, \underline{\mathbf{k}}_o$.

We may assign physical meaning to these operations as follows. The components of the gradient ∇V at any location are the rates at which V is changing with respect to the components of distance in each of these directions. The divergence of ∇V, *i.e.*, $\nabla^2 V$, measures the presence of sources which have produced V and created its field, or flow. For example, when there are no sources for the force in a region, $\nabla^2 V$ is zero throughout that region, but it is a density function over the volume of a physical source of V. The curl, $\underline{\mathbf{B}} = \nabla \times \underline{\mathbf{A}}$, measures the *circulation* or rotation of the vector $\underline{\mathbf{A}}$ around a closed curve in the region containing $\underline{\mathbf{A}}$. It is used to describe the flow of a rotating fluid, or in our case, the presence of a non central force. There are many applications for operators and our list is only a short one. In spherical polar coordinates, these operators take the forms:

(1-1) $$\nabla V = \underline{\hat{\mathbf{u}}}_r\, \partial V/\partial r + \underline{\hat{\mathbf{u}}}_\varphi\, (\partial V/\partial \varphi)/r + \underline{\hat{\mathbf{u}}}_\theta\, (\partial V/\partial \theta)/(r \sin\varphi),$$

(1-2) $$\nabla \bullet \underline{\mathbf{F}} = (1/r^2)\, \partial/\partial r\, (r^2 F_r) + [\partial(F_\varphi \sin\varphi)/\partial \varphi]/(r \sin\varphi)] + [\partial F_\theta/\partial \theta]/(r \sin\varphi),$$

(1-3) $$\nabla \times \underline{\mathbf{A}} = [1/(r \sin\varphi)]\, [\partial/\partial \varphi\, (A_\theta \sin\varphi) - \partial A_\varphi/\partial \theta]\, \underline{\hat{\mathbf{u}}}_r + [1/(r \sin\varphi)]\, [\partial A_r/\partial \theta - (\sin\varphi)\, \partial/\partial r\, (r\, A_\theta)]\, \underline{\hat{\mathbf{u}}}_\varphi + [1/r]\, [\partial/\partial r\, (r\, A_\varphi) - \partial A_r/\partial \varphi]\, \underline{\hat{\mathbf{u}}}_\theta.$$

APPENDIX 2 – ORBITS

The Classical Orbit Solution

Expressions 2-1a and 2-1b specify the position $\underline{\mathbf{r}}$ and the velocity $\underline{\mathbf{v}}$ of a moving object in body-based spherical polar coordinates. Let us further define the object's *specific orbital angular momentum* of $\underline{\mathbf{J}}_{orb} = \underline{\mathbf{r}} \times \underline{\mathbf{v}}$ by equation 2-1c, where the *orbital angular momentum* is $\mu\underline{\mathbf{J}}_{orb}$ for a planet of mass μ. Using Newton's law of gravitation expressed in this coordinate system, the solar force is given to first order by $\underline{\mathbf{F}} = -\mu M g_o\, \underline{\hat{\mathbf{u}}}_r/r^2$, where M is the mass of the sun and g_o is the universal gravitational constant. Applying the force to Newton's acceleration law $\underline{\mathbf{F}} = \mu\, d^2\underline{\mathbf{r}}/dt^2$, we obtain equation 2-2, where the computation of $d^2\underline{\mathbf{r}}/dt^2$ is not trivial. Equations 2-2a, b, and c are its r, φ, and θ components, and they provide the *equations of motion.*

(2-1a) $\underline{\mathbf{r}} = r\,\underline{\hat{\mathbf{u}}}_r,$

(2-1b) $\underline{\mathbf{v}} = \underline{\hat{\mathbf{u}}}_r\, dr_n/dt + r\,(d\varphi/dt)\,\underline{\hat{\mathbf{u}}}_\varphi + r\,(d\theta/dt)\sin\varphi\,\underline{\hat{\mathbf{u}}}_\theta,$

(2-1c) $\underline{\mathbf{J}}_{orb} = \underline{\mathbf{r}} \times \underline{\mathbf{v}} = -\,\underline{\hat{\mathbf{u}}}_\varphi\, r^2 \sin\varphi\, d\theta/dt + \underline{\hat{\mathbf{u}}}_\theta\, r^2\, d\varphi/dt.$

(2-2) $\mu\, d^2\underline{\mathbf{r}}/dt^2 = \underline{\mathbf{F}} = -\,\mu M g_o\,\underline{\hat{\mathbf{u}}}_r/r^2,$

(2-2a) $\mu\, d^2r/dt^2 - \mu\,[r^4\,(d\varphi/dt)^2 + r^4\,(\sin^2\varphi)\,(d\theta/dt)^2]/r^3 = -\,\mu M g_o/r^2,$

(2-2b) $(\mu/r)\,[d/dt\,(r^2\, d\varphi/dt) - r^2\,(d\theta/dt)^2 \sin\varphi\,\cos\varphi] = 0,$

(2-2c) $[\mu/(r\sin\varphi)]\; d/dt\,(r^2 \sin^2\varphi\; d\theta/dt) = 0.$

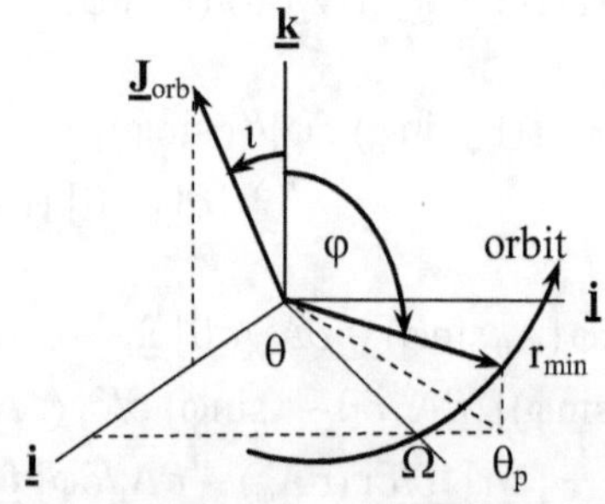

Figure A2.1. Inclined Orbit Geometry

Equation 2-2c is solved by 2-4c, where ι is the inclination angle between $\underline{\mathbf{J}}_{orb}$ and the $\underline{\mathbf{k}}$ axis depicted in Figure A2.1, and the frame has been chosen with $\underline{\mathbf{J}}_{orb}$ lying in the $\underline{\mathbf{ik}}$ plane such that $\cos\varphi = -\sin\iota \cos\theta$. When we express $r^2 (d\theta/dt)^2 \sin\varphi \cos\varphi$ as $-[(J_{orb}{}^2 \cos^2\iota)/(2r^2\, d\varphi/dt)]\, d/dt\, (1/\sin^2\varphi)$ and the term $d/dt\, (r^2\, d\varphi/dt)$ as $[1/(2r^2\, d\varphi/dt)]\, d/dt\, (r^2\, d\varphi/dt)^2$, equation 2-2b becomes 2-3b, whose integral is equation 2-4b. This result may be used to write $r^2\, d\varphi/dt$ as $J_{orb}\, (1 - \cos^2\iota/\sin^2\varphi)^{1/2}$, which is zero at $\varphi = \pi/2 \pm \iota$ and $J_{orb} \sin\iota$ at $\varphi = \pi/2$. †

(2-4c) $r^2\, d\theta/dt = J_{orb}\, (\cos\iota)/(\sin^2\varphi),$

(2-3b) $[1/(2r^3\, d\varphi/dt)]\, d/dt\, [(r^2\, d\varphi/dt)^2 + (J_{orb}{}^2 \cos^2\iota)/(\sin^2\varphi)] = 0,$

(2-4b) $r^4\, (d\varphi/dt)^2 + r^4\, (\sin^2\varphi)\, (d\theta/dt)^2 = |\underline{\mathbf{J}}_{orb}|^2 = J_{orb}{}^2 = \text{constant}.$

Taking φ to be the independent variable, writing dr/dt as $(d\varphi/dt)\, dr/d\varphi$, and substituting $r = 1/u$ into equation 2-2a, we obtain the first and second forms of 2-3a when we input $r^2\, d\varphi/dt = J_{orb}\, (1 - \cos^2\iota/\sin^2\varphi)^{1/2}$. Since $du/d\theta = -(1 - \cos^2\iota/\sin^2\varphi)^{1/2}\, du/d\varphi$, equation 2-3a takes the end form. Expression 2-4a is the solution for r, where θ_p and ε are integration constants.

(2-3a) $(1 - \cos^2\iota/\sin^2\varphi)\, d^2u/d\varphi^2 + [(\cos^2\iota \cos\varphi)/\sin^3\varphi]\, du/d\varphi + u =$
$(1 - \cos^2\iota/\sin^2\varphi)^{1/2}\, d/d\varphi\, [(1 - \cos^2\iota/\sin^2\varphi)^{1/2}\, (du/d\varphi)] + u =$
$d^2u/d\theta^2 + u = Mg_o/J_{orb}{}^2,$ where $\cos\varphi = -\sin\iota \cos\theta,$

(2-4a) $r = 1/u = [J_{orb}{}^2/(Mg_o)]/[1 + \varepsilon \cos(\theta - \theta_p)],$ an ellipse for $0 \le \varepsilon \le 1,$
$r = 1/u = [J^2/(Mg_o)]/[1 + \varepsilon \cos(\theta - \theta_p)],$ a hyperbola for $\varepsilon > 1,$

(2-5a) $J_{orb}{}^2/(Mg_o) = a_n\, (1 - \varepsilon^2),$ for an ellipse,
(2-5b) $J^2/(Mg_o) = a_n\, (\varepsilon^2 - 1),$ for a hyperbola,

(2-6a) $2EJ_{orb}{}^2/(\mu M^2 g_o{}^2) = -1 + \varepsilon^2,$ where E is negative for an ellipse,
(2-6b) $2EJ^2/(\mu M^2 g_o{}^2) = \varepsilon^2 - 1,$ where E is positive for a hyperbola.

---

† If $\underline{\mathbf{J}}_{orb}$ does not lie in the $\underline{\mathbf{ik}}$ plane of the $\underline{\mathbf{ijk}}$ frame, the angular relationships will change. For example, we shall have $\cos\varphi = -\sin\iota \sin\theta$ when $\underline{\mathbf{J}}_{orb}$ lies in the $\underline{\mathbf{jk}}$ plane. However, the general form of the solution for r will be the same.

Whenever $0 \leq \varepsilon \leq 1$, we may write $J_{orb}{}^2/(Mg_o)$ as $a_n(1 - \varepsilon^2)$ in expression 2-5a, specifying an ellipse of eccentricity ε and semi-major axis a_n. The minimum value of $r = r_{min}$ at θ_p is the orbit perihelion depicted in Figure A2.1, which can occur at any angle and is generally different from zero. When ε is greater than 1, the solution is a hyperbola indicated by expression 2-5b, and we write $\underline{\mathbf{r}} \times \underline{\mathbf{v}}$ as $\underline{\mathbf{J}}$ in expression 2-1c to distinguish it from $\underline{\mathbf{J}}_{orb}$ for an ellipse. In both cases it can be shown that ε and the total energy are E are related according to expression 2-6a or b for a given value of J^2 or $J_{orb}{}^2$. When E is negative, the object is orbiting around an attracting center, but when E is positive, it is on an escape trajectory and is never coming back.

Orbits In The Atomic Field

The above derivations may also be applied to orbits in the electromagnetic field of atoms, wherein electrons with negative electrical charges are bound in orbits about a positively charged nucleus. The attractive force is expressed as $-\nabla V_e(r) = -Ze^2\,\underline{\hat{\mathbf{u}}}_r/(4\pi\varepsilon_o r^2)$, where ∇ is the gradient operator, Z is the number of protons in the nucleus, $\underline{\hat{\mathbf{u}}}_r$ is a unit vector in the direction of $\underline{\mathbf{r}}$, and e is the magnitude of charge for the proton and the oppositely charged electron, both of whose approximate values are 1.602×10^{-19} coulomb. The value of $1/(4\pi\varepsilon_o)$ is about 8.9874×10^9 newton-meter-squared per coulomb-squared, and $r = |\underline{\mathbf{r}}|$ is, of course, the separation between the nucleus and the electron. For the one-electron hydrogen atom, the orbit radius is of the order of 10^{-11} meter, which is quite a step from solar gravitational field radii of the order 10^8 kilometers.

Unlike solar orbits, atomic orbits cannot be observed visually and orbit parameters such as ascending nodes and mean anomaly are not relevant to our observation capabilities. Nevertheless, we know that our basic models for the orbit radii, frequencies, and certain other parameters are correct since they provide results that agree with observations of the behaviors of atoms.

APPENDIX 3 – THE MOON'S ORBIT AND ANGULAR MOMENTUM

Let us treat the sum of the orbital angular momenta J_{orb} for the Earth and J_m for the Moon as a single vector $\underline{\mathbf{L}} = \mu_e J_{orb}\, \underline{\mathbf{k}}_c + \mu_m J_m\, \underline{\mathbf{k}}_m$, where $\underline{\mathbf{k}}_c$ is the direction of the Earth's orbit in an $\underline{\mathbf{i}}_c\underline{\mathbf{j}}_c\underline{\mathbf{k}}_c$ frame, and $\underline{\mathbf{J}}_m$ for the Moon lies along its own $\underline{\mathbf{k}}_m$ body axis. The Earth's body mass is μ_e, and the Moon's mass is $\mu_m \cong \mu_e/(81)$. Although $\underline{\mathbf{k}}_c$ is aligned with the Earth's orbit axis, the $\underline{\mathbf{i}}_c\underline{\mathbf{j}}_c\underline{\mathbf{k}}_c$ frame is not the Earth's $\underline{\mathbf{i}}_e\underline{\mathbf{j}}_e\underline{\mathbf{k}}_e$ *ecliptic* frame defined below, but is formed by crossing $\underline{\mathbf{k}}_m$ into $\underline{\mathbf{k}}_c$, where ν_m is the angle between the two vectors. Thus, $\underline{\mathbf{k}}_m \times \underline{\mathbf{k}}_c$ sets both the $\underline{\mathbf{i}}_c$ axis and, initially, the Moon's $\underline{\mathbf{i}}_m$ axis depicted in Figure A3.1. The remaining axes are defined by $\underline{\mathbf{j}}_c = \underline{\mathbf{k}}_c \times \underline{\mathbf{i}}_c$ and $\underline{\mathbf{j}}_m = \underline{\mathbf{k}}_m \times \underline{\mathbf{i}}_m$. The $\underline{\mathbf{i}}_m\underline{\mathbf{j}}_m\underline{\mathbf{k}}_m$ frame is called the Moon's body frame, which is not to be confused with the Moon's actual body.

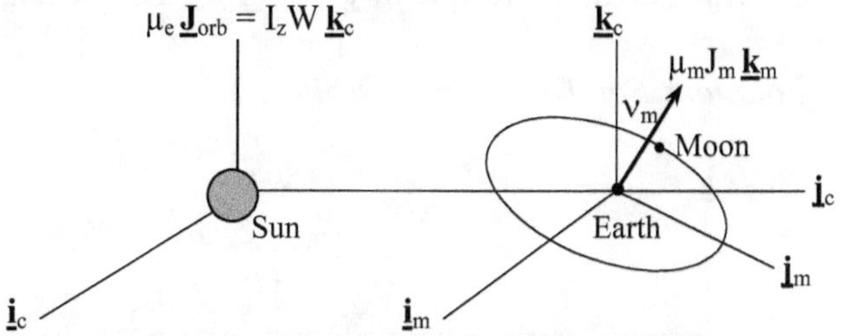

Figure A3.1. Orbits Of The Earth And Moon

Let us express $\mu_e J_{orb}$ as $I_z W = \mu_e a_e^2\, W$, where W is the Earth's average orbital frequency and a_e is its mean orbit radius. Similarly, $\mu_m J_m = \mu_m h^2\, w_m$, where w_m is the Moon's average orbit frequency for a period of 27.32 days and its mean orbit radius is $h \cong 384{,}400$ kilometers. † The lunar body moments are $I_{mz} = I_m = \mu_m h^2$ and $I_{mx} = I_{my} = I_m/2 = I_1$, and the Moon's angular momentum is $\mu_m\, \underline{\mathbf{J}}_m = I_1 \varpi_x\, \underline{\mathbf{i}}_m + I_1 \varpi_y\, \underline{\mathbf{j}}_m + I_m(w_m + \varpi_z)\, \underline{\mathbf{k}}_m$, where $\underline{\boldsymbol{\varpi}}$ is a frame rotation frequency due to the presence of a torque. Expression 3-1 specifies the orbital angular momentum $\underline{\mathbf{L}}_{orb}$ in the $\underline{\mathbf{i}}_c\underline{\mathbf{j}}_c\underline{\mathbf{k}}_c$ frame, where we are ignoring the Earth's spin angular momentum, whose value is about one-fifth of $\mu_m J_m$.

---

† Parameters for the Moon were taken from Bate, R., Mueller, D., and White, J., *Fundamentals of Astrodynamics*, Dover Pub. Inc., NY (1971), at pp. 322-327.

(3-1) $\underline{\mathbf{L}}_{orb} = \mu_m J_m \sin\nu_m \underline{\mathbf{i}}_c + (I_z W + \mu_m J_m \cos\nu_m) \underline{\mathbf{k}}_c$.

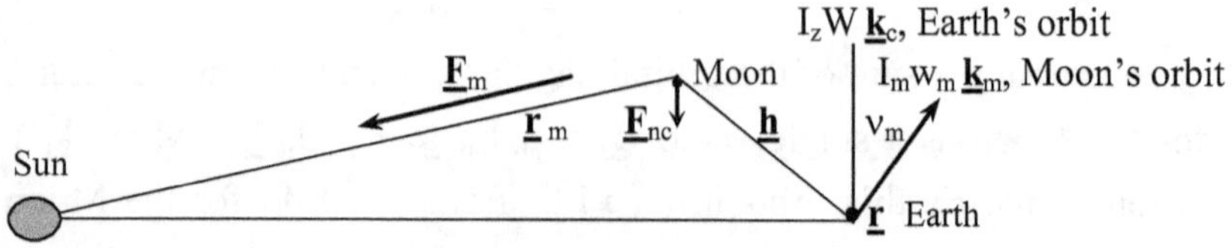

Figure A3.2. Solar Force On The Moon (Exaggerated)

Because the Moon's orbit is slaved to the Earth, as indicated in Figure A3.1, a small torque exists due to an effective non-central force in the solar attraction. More specifically, from Newton's law the solar force on the Moon is $\underline{\mathbf{F}}_m = -\mu_m M g_o \underline{\mathbf{r}}_m / r_m^3$, where $\underline{\mathbf{r}}_m$ is the radial vector from the sun to the Moon depicted in Figure A3.2. Since the Moon is not collocated with the Earth at $\underline{\mathbf{r}}$, the force $\underline{\mathbf{F}}_m$ includes a small component $\underline{\mathbf{F}}_{nc}$ normal to the Moon's orbit, which creates a torque of $\underline{\boldsymbol{\tau}} = \underline{\mathbf{r}} \times \underline{\mathbf{F}}_{nc}$ on the Moon's orbit.

Frame Rotations For The Moon's Orbit

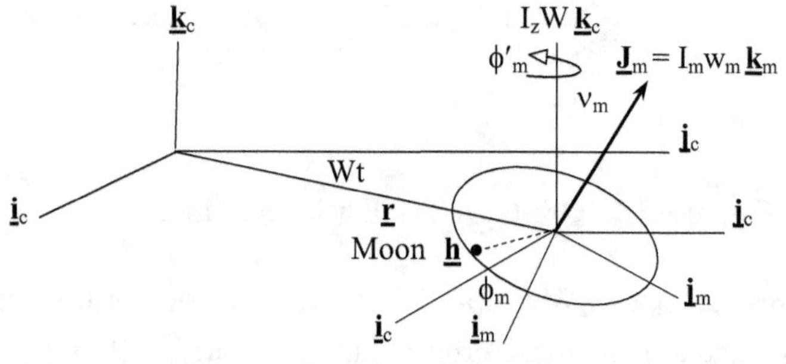

Figure A3.3. Orbit-Based Reference Frame For The Moon

By deriving an expression for the torque, we can compute $d\underline{\mathbf{L}}/dt + \underline{\boldsymbol{\omega}} \times \underline{\mathbf{L}} = \underline{\boldsymbol{\tau}}$ to obtain $\underline{\mathbf{L}}$, the total angular momentum including frame rotations. Let us specify variations in the angles ν_m between $\underline{\mathbf{J}}_m$ and $\underline{\mathbf{k}}_c$, and in ϕ_m between $\underline{\mathbf{i}}_c$ and $\underline{\mathbf{i}}_m$, as indicated in Figure A3.3. The variable ν'_m is the equivalent of nutation for a spin axis, and ϕ'_m is the precession of the Moon's orbit. † (If

† Compare with the oblate Earth frame rotation found in Goldstein, H., *Classical Mechanics* (1980 ed.), §5-8, pp. 226-231, Addison-Wesley Pub. Co., The Philippines.

we fail to model frame rotations for both precession ϕ_m and nutation ν_m, we would be unable to derive the variable inclination of the Moon's orbit.)

Starting from the Earth's orbit, we reach the lunar frame by performing a right-handed rotation of ϕ_m about $\underline{\mathbf{k}}_c$, and then rotating by ν_m about $\underline{\mathbf{i}}_m$. A column vector $\underline{\boldsymbol{\tau}}_c$ in the Earth's frame becomes $\underline{\boldsymbol{\tau}}_m = \underline{\boldsymbol{C}}_{\nu\phi}\, \underline{\boldsymbol{\tau}}_c$ in the lunar frame, as specified by matrix expression 3-2. The Earth's location in the $\underline{\mathbf{i}}_c\underline{\mathbf{j}}_c\underline{\mathbf{k}}_c$ frame is $\underline{\mathbf{r}} = a_e\,(\cos Wt\, \underline{\mathbf{i}}_c + \sin Wt\, \underline{\mathbf{j}}_c)$, where W = 2π radians per 365.25 days. The Moon's initial location is specified by $\underline{\mathbf{h}} = h\,(-\sin w_m t\, \underline{\mathbf{i}}_m + \cos w_m t\, \underline{\mathbf{j}}_m)$ in its own frame, where w_m is 2π radians per 27.32 days. Applying the matrix product $\underline{\boldsymbol{C}}_{\nu\phi}^{-1} = \underline{\boldsymbol{C}}_{\phi}^{-1}\, \underline{\boldsymbol{C}}_{\nu}^{-1}$ to $\underline{\mathbf{h}}$, the Moon's location in the $\underline{\mathbf{i}}_c\underline{\mathbf{j}}_c\underline{\mathbf{k}}_c$ frame will be $\underline{\mathbf{r}} + \underline{\mathbf{h}}$, where $\underline{\mathbf{h}}$ and $|\underline{\mathbf{r}} + \underline{\mathbf{h}}|^2$ are specified by expressions 3-3 and 3-4. †

$$(3\text{-}2)\qquad \underline{\boldsymbol{C}}_{\nu\phi} = \underline{\boldsymbol{C}}_{\nu}\,\underline{\boldsymbol{C}}_{\phi} = \begin{bmatrix} 1 & 0 & 0 \\ 0 & \cos\nu_m & \sin\nu_m \\ 0 & -\sin\nu_m & \cos\nu_m \end{bmatrix} \begin{bmatrix} \cos\phi_m & \sin\phi_m & 0 \\ -\sin\phi_m & \cos\phi_m & 0 \\ 0 & 0 & 1 \end{bmatrix},$$

$$(3\text{-}3)\qquad \underline{\mathbf{h}} = -h\,[(\sin w_m t \cos\phi_m + \cos\nu_m \cos w_m t \sin\phi_m)\,\underline{\mathbf{i}}_c + (\sin w_m t \sin\phi_m - \cos\nu_m \cos w_m t \cos\phi_m)\,\underline{\mathbf{j}}_c - \sin\nu_m \cos w_m t\, \underline{\mathbf{k}}_c],$$

$$(3\text{-}4)\qquad |\underline{\mathbf{r}} + \underline{\mathbf{h}}|^2 = a_e^2 - 2a_e h\,[\cos Wt\,(\sin w_m t \cos\phi_m + \cos w_m t \cos\nu_m \sin\phi_m) + \sin Wt\,(\sin w_m t \sin\phi_m - \cos w_m t \cos\nu_m \cos\phi_m)] + h^2\,(\ldots)$$
$$\cong a_e^2 + 2a_e h\,[-\sin w_m t \cos(Wt - \phi_m) + \cos w_m t \cos\nu_m \sin(Wt - \phi_m)],$$

$$(3\text{-}5)\qquad 1/|\underline{\mathbf{r}} + \underline{\mathbf{h}}|^3 \cong \{\,1 + 3\,[h/a_e]\,[\sin w_m t \cos(Wt - \phi_m) - \cos\nu_m \cos w_m t \sin(Wt - \phi_m)]\,\}/a_e^3.$$

Approximating $1/|\underline{\mathbf{r}} + \underline{\mathbf{h}}|^3$ to first order by expression 3-5, the force component $\underline{\mathbf{F}}_h$ aligned with $\underline{\mathbf{h}}$ is given by expression 3-6.

$$(3\text{-}6)\qquad \underline{\mathbf{F}}_h = -\mu_m M g_o\, \underline{\mathbf{h}}/|\underline{\mathbf{r}} + \underline{\mathbf{h}}|^3$$
$$= \{\mu_m M g_o h/a_e^3\}\,\{1 + [3h/a_e]\,[\sin w_m t \cos(Wt - \phi_m) - \cos\nu_m \cos w_m t \sin(Wt - \phi_m)]\}$$
$$\{(\sin w_m t \cos\phi_m + \cos w_m t \cos\nu_m \sin\phi_m)\,\underline{\mathbf{i}}_c + (\sin w_m t \sin\phi_m - \cos w_m t \cos\nu_m \cos\phi_m)\,\underline{\mathbf{j}}_c - \sin\nu_m \cos w_m t\, \underline{\mathbf{k}}_c\},$$

---

† The inverse of $\underline{\boldsymbol{C}}_{\nu\phi}$, *i.e.*, $\underline{\boldsymbol{C}}_{\phi}^{-1}\underline{\boldsymbol{C}}_{\nu}^{-1}$, is its transpose since $\underline{\boldsymbol{C}}_{\nu\phi}$ is a unitary matrix. We have specified the vectors in column form for multiplication by 3 × 3 matrices.

$$(3\text{-}7)\quad \underline{\mathbf{r}} \times \underline{\mathbf{F}}_{nc} = \{\mu_m M g_o h/a_e^2\}\,\{[\sin\nu_m \cos w_m t]\,[-\sin Wt\,\underline{\mathbf{i}}_c + \cos Wt\,\underline{\mathbf{j}}_c] - [\cos w_m t \cos\nu_m \cos(Wt-\phi) + \sin w_m t \sin(Wt-\phi_m)]\,\underline{\mathbf{k}}_c\} + \{3\mu_m h^2 M g_o/a_e^3\}\,\{[\cos^2 w_m t \sin\nu_m \cos\nu_m \sin(Wt-\phi_m) - \sin w_m t \cos w_m t \sin\nu_m \cos(Wt-\phi_m)][\sin Wt\,\underline{\mathbf{i}}_c - \cos Wt\,\underline{\mathbf{j}}_c] + [(\cos^2 w_m t \cos^2\nu_m - \sin^2 w_m t)\sin(Wt-\phi_m)\cos(Wt-\phi_m) + \sin w_m t \cos w_m t \cos\nu_m (1-2\cos^2(Wt-\phi_m))]\,\underline{\mathbf{k}}_c\}.$$

Since $\underline{\mathbf{r}} \times \underline{\mathbf{r}} \equiv 0$, expression 3-7 specifies the torque $\underline{\boldsymbol{\tau}}_c = \underline{\mathbf{r}} \times \underline{\mathbf{F}}_{nc}$ in the $\underline{\mathbf{i}}_c\underline{\mathbf{j}}_c\underline{\mathbf{k}}_c$ frame. Using Kepler's third law, we may set $Mg_o/a_e^3 = W^2$. The average values of $\sin^2 w_m t$ and $\cos^2 w_m t$ are 1/2, † and averaging to zero over the orbit the terms not multiplied by $\sin^2 w_m t$ or $\cos^2 w_m t$, we obtain equation 3-8 for $\underline{\boldsymbol{\tau}}_c$ in the $\underline{\mathbf{i}}_c\underline{\mathbf{j}}_c\underline{\mathbf{k}}_c$ frame. Applying $\underline{\boldsymbol{C}}_{\nu\phi}$ to $\underline{\boldsymbol{\tau}}_c$, the $\underline{\mathbf{k}}_m$ component in the Moon frame vanishes, and equation 3-9 provides the torque $\underline{\boldsymbol{\tau}}_m$ in this frame.

$$(3\text{-}8)\quad \underline{\boldsymbol{\tau}}_c = [\underline{\mathbf{r}} \times \underline{\mathbf{F}}]_{w\text{-}av} = (3/2)\,I_m W^2\,[\sin\nu_m \sin(Wt-\phi_m)]\,[\cos\nu_m \sin Wt\,\underline{\mathbf{i}}_c - \cos\nu_m \cos Wt\,\underline{\mathbf{j}}_c - \sin\nu_m \cos(Wt-\phi_m)\,\underline{\mathbf{k}}_c],$$

$$(3\text{-}9)\quad \underline{\boldsymbol{\tau}}_m = \underline{\boldsymbol{C}}_{\nu\phi}\,[\underline{\mathbf{r}} \times \underline{\mathbf{F}}]_{w\text{-}av} = (3/2)\,I_m W^2\,[\sin\nu_m \cos\nu_m \sin^2(Wt-\phi_m)\,\underline{\mathbf{i}}_m - \sin\nu_m \sin(Wt-\phi_m)\cos(Wt-\phi_m)\,\underline{\mathbf{j}}_m].$$

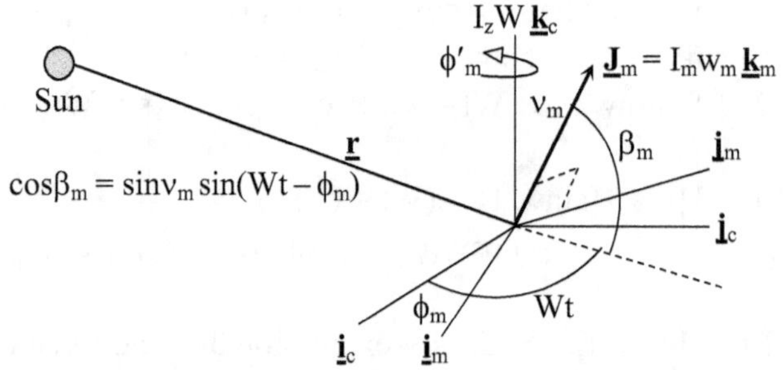

Figure A3.4. Body Angles Used To Specify Lunar Frame Rotations

Our results define an angle β_m between $\underline{\mathbf{J}}_m$ and $\underline{\mathbf{r}}$ given by $\cos\beta_m = \underline{\mathbf{k}}_m \bullet \underline{\hat{\mathbf{u}}}_r = \sin\nu_m \sin(Wt-\phi_m)$ in Figure A3.4, where $\underline{\mathbf{k}}_m = \sin\nu_m(-\sin\phi_m\,\underline{\mathbf{i}}_c + \cos\phi_m\,\underline{\mathbf{j}}_c)$ and $\underline{\hat{\mathbf{u}}}_r = \underline{\mathbf{r}}/a_e = \cos Wt\,\underline{\mathbf{i}}_c + \sin Wt\,\underline{\mathbf{j}}_c$. The form of $Wt - \phi_m$ indicates that ϕ_m subtracts from Wt to produce β_m. In a frame rotating with ϕ_m the variables appear as $\omega_x = d\nu_m/dt = \nu'_m$, $\omega_y = 0$, and $\omega_z = d\phi_m/dt = \phi'_m$.

---

† If we had selected a phase angle ≠ 0, the effects would disappear upon averaging.

Applying $\underline{C}_v$ to these forms, they become $\varpi_x = \nu'_m$, $\varpi_y = \phi'_m \sin\nu_m$, and $\varpi_z = \phi'_m \cos\nu_m$ in the Moon's frame. The total angular momentum $\underline{\mathbf{L}}$ for our Earth and Moon system model then becomes expression 3-10. Setting $d\underline{\mathbf{L}}/dt$ to $\underline{\tau}_m$, the Euler equations discussed in the text above take the forms of 3-11a, b, and c, where all of the terms multiplied by $I_z W$ cancel.

$$\text{(3-10)} \qquad \underline{\mathbf{L}} = I_1\, \varpi_x \underline{\mathbf{i}}_m + [I_1\, \varpi_y + I_z W \sin\nu_m]\, \underline{\mathbf{j}}_m + [I_m (w_m + \varpi_z) + I_z W \cos\nu_m]\, \underline{\mathbf{k}}_m,$$

$$\text{(3-11a)} \qquad I_1\, d\varpi_x/dt + [I_m (w_m + \varpi_z) - I_1\, \varpi_z]\, \varpi_y = (3/2)\,(I_m - I_1)\, W^2 (\sin\nu_m \cos\nu_m)\,[1 - \cos 2(Wt - \phi_m)],$$

$$\text{(3-11b)} \qquad I_1\, d\varpi_y/dt - [I_m(w_m + \varpi_z) - I_1 \varpi_z]\, \varpi_x = -(3/2)\,(I_m - I_1)\, W^2 \sin\nu_m \sin 2(Wt - \phi_m),$$

$$\text{(3-11c)} \qquad I_m\, d/dt\,(w_m + \varpi_z) = 0, \qquad\qquad w_m + \varpi_z = w_o, \text{ a constant.}$$

If w_m is much greater than ϖ_z and $I_m w_m \varpi_x$ is much greater than $I_1\, d\varpi_y/dt$, and if $\phi_m = \phi'_m t$ is much less than Wt, as observations indicate to be true, equation 3-11b becomes 3-12. Integration provides equation 3-13 for ν_m, which shows that ν_m oscillates about its median angle ν_{mo} at a frequency of twice the orbital value, with the extremes of ν_m set by $\exp[\pm 3W/(8w_m)]$, where "exp" is the base of natural logarithms. Using average orbital periods of 27.32 days and 365.25 days for the Moon and the Earth, respectively, the exponential factor is 0.028. Setting ν_{mo} to the observed median value for ν_m of 5° 8′, the lunar inclination extremes become 5° 17′ and 4° 59′ in our model, in good agreement with the observed values of 5° 18′ and 4° 59′.

$$\text{(3-12)} \qquad \varpi_x = d\nu_m/dt \cong [3W^2/(4w_m)]\, \sin\nu_m \sin 2Wt,$$

$$\text{(3-13)} \qquad \tan(\nu_m/2) = [\tan(\nu_{mo}/2)]\, \exp[-3W\,(\cos 2Wt)/(8w_m)].$$

When we substitute $\varpi_y = \phi'_m \sin\nu_m$, with $I_m w_m \varpi_y$ being much greater than $I_1\, d\varpi_x/dt$, equation 3-11a provides equation 3-14 for ϕ'_m.

$$\text{(3-14)} \qquad \phi'_m \cong [3W^2/(4w_m)]\,(\cos\nu_m)\,(1 - \cos 2Wt).$$

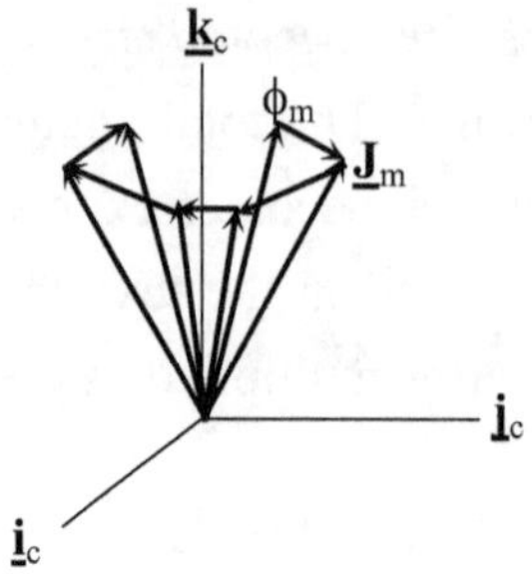

Figure A3.5. Rotation Of The Moon's Orbital Vector

The oscillatory component of equation 3-14 averages to zero over half of the Earth's orbital period, but the secular component produces almost constant precession of $\underline{\mathbf{J}}_m$ about $\underline{\mathbf{k}}_e$ at a rate of $\phi'_m = 3W^2(\cos\nu_m)/(4w_m)$, as roughly sketched in Figure A3.5. Using $\nu_{mo} = 5.1$ degrees as the average of ν_m, we obtain a value of about 0.056W for ϕ'_m. Our result is in reason agreement with the observed average westward rotation of the Moon's orbit vector over a period of 18.6 years, *i.e.*, $\phi'_m \cong 0.054W$, and would be improved by including the torque due to the Earth's oblateness. We note that the term $I_m\phi'_m \cos\nu_m$ is about 0.4 percent of the Moon's total angular momentum.

Since $\varpi_z = \phi'_m \cos\nu_m$ is not constant, equation 3-11c requires w_m to vary as ϖ_z varies. The median value of $\phi'_m \cos\nu_m \cong 0.0042\, w_m$ represents about 2.75 hours of the average period of $t_m \cong 655.7$ hours for w_m. Multiplying equation 3-14 by $\cos\nu_m$ we find that $\phi'_m \cos\nu_m$ varies by a factor of about twice its median value over a time span of approximately 183 days, indicating a variation in the period of about 5.5 hours. The period for the Moon's orbit is observed to vary as much as seven hours from its average. Since the period is expressible as $2\pi h^2(1-\varepsilon^2)^{1/2}/J_m$, the orbit eccentricity ε must vary with the period, which is called the *lunar evection*.

A Third Rotation Component And The Lunar Lagrangian

Limiting frame rotations to ν'_m and ϕ'_m does not allow for a third frame rotation, or degree of freedom, which Euler included in his formulations. Some authors incorrectly treat body spin as a third variable. Unlike body spin, frame rotation about a third axis multiplies <u>all</u> of the inertial moments.

We shall model this component by the variable ξ'_m, illustrated in Figure A3.6. The $\underline{C}_\xi$ matrix, expression 3-15, rotates a column vector about $\underline{\mathbf{k}}_m$ to complement $\underline{C}_{\nu\phi}$. Applying $\underline{C}_\xi$ to $\underline{\varpi}$, we obtain a vector $\underline{\upsilon}$ which models rotations in an inertial frame aligned with the initial lunar frame. Expression 3-16 specifies $\underline{\upsilon}$ and the other rotation vectors that include ξ'_m.

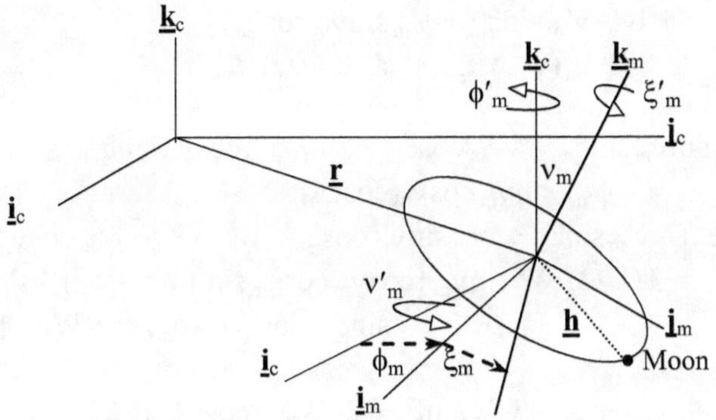

Figure A3.6. A Complete Set Of Frame Rotations

$$(3\text{-}15)\quad \underline{C}_\xi = \begin{bmatrix} \cos\xi & \sin\xi & 0 \\ -\sin\xi & \cos\xi & 0 \\ 0 & 0 & 1 \end{bmatrix},$$

$$(3\text{-}16)\quad \underline{\mathbf{w}}:\ \begin{aligned} w_x &= \nu'\cos\phi + \xi'\sin\nu\sin\phi \\ w_y &= \nu'\sin\phi - \xi'\sin\nu\cos\phi \\ w_z &= \phi' + \xi'\cos\nu, \end{aligned} \qquad \underline{\omega}:\ \begin{aligned} \omega_x &= \nu' \\ \omega_y &= -\xi'\sin\nu \\ \omega_z &= \phi' + \xi'\cos\nu, \end{aligned}$$

$$\underline{\varpi}:\ \begin{aligned} \varpi_x &= \nu' \\ \varpi_y &= \phi'\sin\nu \\ \varpi_z &= \xi' + \phi'\cos\nu, \end{aligned} \qquad \underline{\upsilon}:\ \begin{aligned} \upsilon_x &= \nu'\cos\xi + \phi'\sin\nu\sin\xi \\ \upsilon_y &= -\nu'\sin\xi + \phi'\sin\nu\cos\xi \\ \upsilon_z &= \xi' + \phi'\cos\nu. \end{aligned}$$

In a frame which is initially aligned with the Earth's inertial frame but is rotating with ϕ, the frame rotation vector appears as $\underline{\omega} = \underline{C}_\nu^{-1}\,\underline{\varpi}$, and in the inertial frame for Earth's orbit it appears as $\underline{\mathbf{w}} = \underline{C}_\phi^{-1}\,\underline{\omega}$. The inertial frame expressions for $\underline{\mathbf{w}}$ and $\underline{\upsilon}$ are forms of *Euler's geometrical relationships.*

Applying $\underline{C}_\xi$ to $\underline{\tau}_m$ where $I_{mx} = I_{my} = I_z/2 = I_1$, the body angular momentum $\mu_m\,\underline{\mathbf{J}}_m$ and the components of $\mu_m\, d\underline{\mathbf{J}}_m/dt + \underline{\upsilon} \times \mu_m\,\underline{\mathbf{J}}_m$ in the end inertial frame are given by expression 3-17 and by expressions 3-18a, b, and c, respectively. † Inclusion of frame rotation of ξ' does not affect the angle $\beta_m = \text{arc-cos}(\underline{\mathbf{k}}_m \bullet \underline{\hat{\mathbf{u}}}_r)$ in Figure 3.4, nor our torque calculation since $\sin^2 w_m t$ is replaced by $\sin^2(w_m t + \xi)$, etc.

(3-17) $$\mu_m \underline{\mathbf{J}}_m = I_1 (v'_m \cos\xi_m + \phi'_m \sin v_m \sin\xi_m)\, \underline{\mathbf{i}}_m + I_1 (-v'_m \sin\xi_m + \phi'_m \sin v_m \cos\xi_m)\, \underline{\mathbf{j}}_m + I_m (w_m + \xi'_m + \phi'_m \cos v_m)\, \underline{\mathbf{k}}_m,$$

(3-18a) $$I_1 (\phi''_m \sin v_m \sin\xi_m + v''_m \cos\xi_m + 2v'_m \phi'_m \cos v_m \sin\xi_m - \phi'^2_m \sin v_m \cos v_m \cos\xi_m) + I_m (-v'_m \sin\xi_m + \phi'_m \sin v_m \cos\xi_m)(w_m + \xi'_m + \phi'_m \cos v_m) = 3 (I_m/2) W^2 \sin v_m [\cos v_m \cos\xi_m \sin^2(Wt - \phi_m) - \sin\xi_m \sin(Wt - \phi_m) \cos(Wt - \phi_m)],$$

(3-18b) $$I_1 (\phi''_m \sin v_m \cos\xi_m - v''_m \sin\xi_m + 2v'_m \phi'_m \cos v_m \cos\xi_m + \phi'^2_m \sin v_m \cos v_m \sin\xi_m) - I_m (v'_m \cos\xi_m + \phi'_m \sin v_m \sin\xi_m)(w_m + \xi'_m + \phi'_m \cos v_m) = -3 (I_m/2) W^2 \sin v_m [\cos v_m \sin\xi_m \sin^2(Wt - \phi_m) + \cos\xi_m \sin(Wt - \phi_m) \cos(Wt - \phi_m)],$$

(3-18c) $$I_m\, d/dt\, (w_m + \xi'_m + \phi'_m \cos v_m) = 0.$$

Next applying $\underline{C}_\xi^{-1}$ to the vector components modeled by equations 3-18a, b, and c, and multiplying the result for 3-18b by $\sin v_m$, we obtain equations 3-19a, b, and c as the components of frame rotation in the end inertial frame. If we define a body potential energy V_b by expression 3-20, the right sides of equations 3-19a, b, and c are $-\partial V_b/\partial v_m$, $-\partial V_b/\partial \phi_m$, and $-\partial V_b/\partial \xi_m$. We have included the "constant" factor $(I_m - I_1)\, W^2/2$ in V_b since the normalized second order Legendre solution for Laplace's equation requires V_b to be of this form as a function of $\cos\beta_m$. The Lagrangian based on V_b, expression 3-21, is amenable to including other energy expressions

† The frame defined by $\underline{\varpi}$ rotates with ξ' and is accelerated. Inputting its components to Euler's equations leads to a wrong result wherein ξ' multiplies I_m and I_1. However, $\underline{\varpi}$ can be used to derive the equations of motion from the rotational energy when $I_x = I_y$ since $I_1 \upsilon_x^2 + I_1 \upsilon_y^2 + I_m (w_m + \upsilon_z)^2 = I_1 \varpi_x^2 + I_1 \varpi_y^2 + I_m (w_m + \varpi_z)^2$.

and simplifies the derivations of the equations of motion at this basic level.

(3-19a) $$I_l\, v''_m + [I_m(w_m + \xi'_m + \phi'_m \cos v_m) - I_l\, \phi'_m \cos v_m]\,(\phi'_m \sin v_m) = 3\,(I_m/2)\, W^2 \sin v_m \cos v_m \sin^2(Wt - \phi_m),$$

(3-19b) $$I_l \sin v_m\, d/dt\,(\phi'_m \sin v_m) - [I_m(w_m + \xi'_m + \phi'_m \cos v_m) - I_l\, \phi'_m \cos v_m]\,(v'_m \sin v_m) = -3\,(I_m/2)\, W^2 \sin^2 v_m \sin(Wt - \phi_m)\cos(Wt - \phi_m),$$

(3-19c) $$I_m\, d/dt\,(w_m + \xi'_m + \phi'_m \cos v_m) = 0,^{\dagger}$$

(3-20) $$V_b = -(I_m - I_l)\, W^2\, [(3/2)\sin^2 v_m \sin^2(Wt - \phi_m) - 1/2] = -(I_m - I_l)\, W^2\, [(3/2)\cos^2\beta_m - 1/2],$$

(3-21) $$L_b = T_b - V_b = I_l\, \upsilon_x^2/2 + I_l\, \upsilon_y^2/2 + I_m(w_m + \upsilon_z)^2/2 - V_b = I_l\,(v'^2_m + \phi'^2_m \sin^2 v_m)/2 + I_m(w_m + \xi'_m + \phi'_m \cos v_m)^2/2 - V_b.$$

Modeling the motion of the Moon has been a difficult problem. Isaac Newton stated to his friend Edmund Halley that analyzing its motion made his head ache and kept him awake so often that he "would think of it no more". Solutions for its motion are especially required for ship navigation, and precision considerably greater than that presented is required. George W. Hill published a more exact theory in 1878, but the navigation requirements were still not met until Ernest W. Brown improved on Hill's work and published his *Tables of the Motion of the Moon* in 1919. The Hill-Brown theory was ultimately replaced in 1984 by the use of methods of space age technology. We have included the above derivation in this presentation to demonstrate the viability of our approach in determining frame rotations for the Earth to a first order using non-specialized coordinate systems.

---

† A form of torque appears in equation 3-19c when a central force field interacts with an asymmetrical mass distribution in the horizontal body plane, *i.e.*, $I_x \neq I_y$.

APPENDIX 4 – VECTOR POTENTIAL EFFECTS ON SOLAR ORBITS

Referring to Figure 4.3 above, let us examine in more detail the effects on an orbit of the vector potential $\underline{\mathbf{A}}_g$ modeled by expression 4-1. Expressing $\underline{\mathbf{v}}$ as $r'\,\underline{\hat{\mathbf{u}}}_r + r\,\varphi'\,\underline{\hat{\mathbf{u}}}_\varphi + r\,\theta'\sin\varphi\,\underline{\hat{\mathbf{u}}}_\theta$ for the orbit of a planet of mass μ_p, we propose expression 4-2 as the base level Lagrangian L_A that includes $\underline{\mathbf{A}}_g$ but is not undergoing frame rotation. Computing $d/dt\,(\partial L_A/\partial q_i') = \partial L_A/\partial q_i$ for the r, φ, and θ coordinates, we obtain equations 4-3a, b, and c.

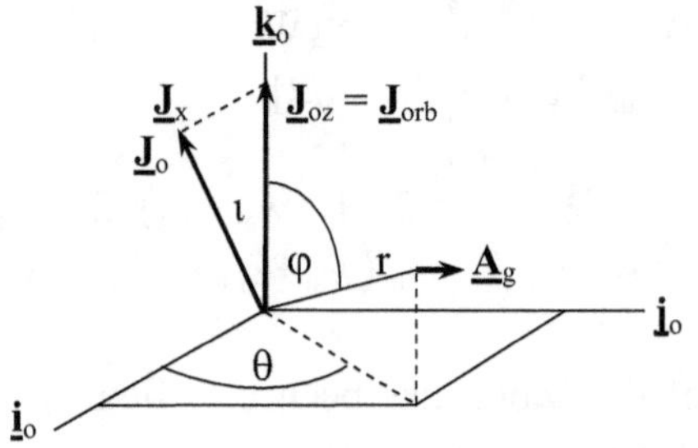

Figure A4.1. The Gravitational Vector Potential $\underline{\mathbf{A}}_g$

(4-1) $\underline{\mathbf{A}}_g = \sigma(\cos\theta\,\underline{\mathbf{i}}_o + \sin\theta\,\underline{\mathbf{j}}_o)/(2r) = \sigma(\sin\varphi\,\underline{\hat{\mathbf{u}}}_r + \cos\varphi\,\underline{\hat{\mathbf{u}}}_\varphi)/(2r),$

(4-2)
$$L_A/\mu_p = (\underline{\mathbf{v}} + \underline{\mathbf{A}}_g)^2/2 - V/\mu_p$$
$$= [(r' + (\sigma/2)(\sin\varphi)/r]^2/2 + [r\,\varphi' + (\sigma/2)(\cos\varphi)/r]^2/2 + (r^2\theta'^2\sin^2\varphi)/2 + Mg_o/r,$$

(4-3a)
$$d/dt\,[r' + (\sigma/2)(\sin\varphi)/r]$$
$$= r\,\theta'^2\sin^2\varphi + [r' + (\sigma/2)(\sin\varphi)/r]\,[-(\sigma/2)(\sin\varphi)/r^2] + [r\,\varphi' + (\sigma/2)(\cos\varphi)/r]\,[\varphi' - (\sigma/2)(\cos\varphi)/r^2] - Mg_o/r^2,$$

(4-3b)
$$d/dt\,[r^2\varphi' + \sigma(\cos\varphi)/2]$$
$$= r^2\theta'^2\sin\varphi\cos\varphi + [r' + (\sigma/2)(\sin\varphi)/r]\,[(\sigma/2)(\cos\varphi)/r] + [r\,\varphi' + (\sigma/2)(\cos\varphi)/r]\,[-(\sigma/2)(\sin\varphi)/r],$$

(4-3c) $d/dt\,(r^2\theta'\sin^2\varphi) = d/dt\,(\partial L_A/\partial\theta') = \partial L_A/\partial\theta = 0,$

(4-4c) $r^2\theta'\sin^2\varphi = J_{oz}.$

Expression 4-4c is the solution for 4-3c. Using this constant J_{oz} to re-write equation 4-3b as 4-4b, all σ terms cancel except for the $\sigma r'$ term, and we obtain expression 4-5b for the constant solution J_o^2. Similarly, equation 4-3a becomes 4-4a when we cancel like terms, and 4-5a is the solution for r.

(4-4b) $d/dt\,(r^4\varphi'^2 + J_{oz}^2/\sin^2\varphi) = d/dt\,(r^4\varphi'^2 + r^4\theta'^2\sin^2\varphi) = \sigma r'\, r\, \varphi' \cos\varphi$,

(4-5b) $r^4\varphi'^2 + r^4\theta'^2\sin^2\varphi - \sigma\int r'\, r\cos\varphi\, d\varphi = J_o^2$,

(4-4a) $d^2r/dt^2 - [r^4\varphi'^2 + r^4\theta'^2\sin^2\varphi - \sigma^2/4 - \sigma r^2\varphi'(\cos\varphi)/2]/r^3 = -Mg_o/r^2$,

(4-5a) $r = 1/u = a_n(1-\varepsilon^2)/[1+\varepsilon\cos(\theta-\theta_p)]$,

where $a_n(1-\varepsilon^2) = J_{orb}^2/(Mg_o)$, and $\cos\varphi = -\sin\iota\cos(\theta-\theta_p)$,

(4-6) $J_{orb}^2 = r^2\varphi'[r^2\varphi' - \sigma(\cos\varphi)/2] + J_{oz}^2/\sin^2\varphi - \sigma^2/4$, where $J_{orb}^2 = J_{oz}^2$.

Inputting J_{oz} from 4-4c, equation 4-4a defines the squared vector J_{orb}^2, expression 4-6, which implies a hyperbolic relationship between its components. Although $\underline{\mathbf{A}}_g$ changes direction with θ, it produces an imaginary vector $\underline{\mathbf{J}}_x$ of square magnitude $\sigma^2/4$ which remains fixed in inertial space and is orthogonal to the real specific angular momentum $\underline{\mathbf{J}}_{oz}$. In the configuration depicted in Figure A4.2, $\underline{\mathbf{J}}_x$ causes the observed orbit vector $\underline{\mathbf{J}}_o$ to be inclined to $\underline{\mathbf{k}}_o$ at the angle ι specified by expression 4-7. $\underline{\mathbf{J}}_x$ combines with $\underline{\mathbf{J}}_o$ hyperbolically to form $\underline{\mathbf{J}}_{oz}$, the projection of specific angular momentum on $\underline{\mathbf{k}}_o$, the field's polar axis. Absent frame rotation, $\underline{\mathbf{J}}_{oz}$ is the orbit vector $\underline{\mathbf{J}}_{orb}$.

(4-7) $\tan\iota = \sigma/(2J_o)$.

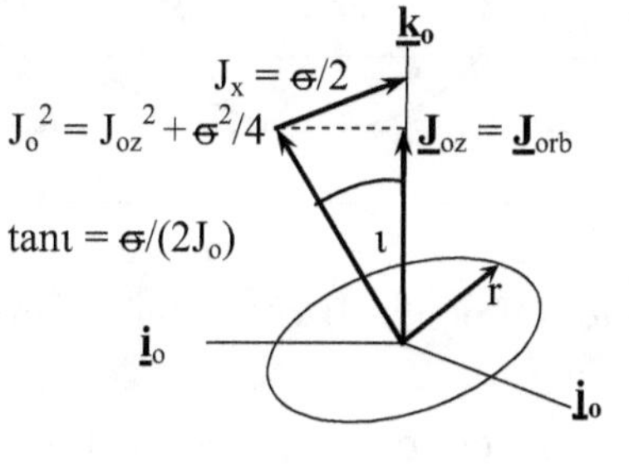

Figure A4.2. Orbit Vector Including $\underline{\mathbf{A}}_g$ Without Frame Rotation

Since r′ is zero for a circular orbit, we shall treat $\varpi \int r' \, r \cos\varphi \, d\varphi$ in equation 4-5b as a small perturbation for planetary orbits with an unperturbed value of $J_o^2 = r^4\varphi'^2 + r^4\theta'^2 \sin^2\varphi = J_{oz}^2 + \varpi^2/4$. The term $-r^2\varphi' \varpi (\cos\varphi)/2$ is a small oscillation about $\varphi = \pi/2$, where $r^4\varphi'^2$ is due solely to $\underline{\mathbf{A}}_g^2$ and becomes $\varpi^2/4$ at $\varphi = \pi/2$. These two ϖ terms are components of a small magnetic-like force $\underline{\mathbf{F}}_B = \mu_p \underline{\mathbf{v}} \times \underline{\mathbf{B}}_g$, where $\underline{\mathbf{B}}_g = \nabla \times \underline{\mathbf{A}}_g$. Absent frame rotations, $\underline{\mathbf{J}}_{orb}$ is $\underline{\mathbf{J}}_{oz}$ along the $\underline{\mathbf{k}}_o$ axis, and the solution for r is unchanged from its solution when $\underline{\mathbf{A}}_g$ is not present. However, the term $\varpi^2/4$ is now included in the observed orbit vector $\underline{\mathbf{J}}_o$, which is constant with the small perturbation $\varpi \int r' \, r \cos\varphi \, d\varphi$.

If the term $-\varpi(\cos\varphi)/2$ were not present in the expression 4-6 for J_{orb}^2, we would have a simple inclined orbit with $r^2\varphi' = 0$ at $\varphi = \pi/2 \pm \iota$. Instead, the term $p^2(\varphi) = r^2\varphi' \, [r^2\varphi' - \varpi(\cos\varphi)/2]$ behaves like the square of a real specific angular momentum component $p(\varphi)$ such that $r^2\varphi'$ must not be less than $\varpi(\cos\varphi)/2$ for positive values of $\cos\varphi$. A like condition holds for negative values of $\cos\varphi$. Thus, $r^2\varphi'$ never becomes zero, and $p(\varphi)$ abruptly reverses its sign when $\theta - \theta_p$ is an integral multiple of π. Inputting $J_{orb}^2 = J_{oz}^2$ to equation 4-6 with $\tan\iota = \varpi/(2J_o)$, we obtain equation 4-8 for $p^2(\varphi)$ at $\varphi = \pi/2 + \iota$.

$$(4\text{-}8) \quad p^2(\pi/2 \pm \iota) = r^2\varphi' \, [r^2\varphi' - \varpi(\sin\iota)/2] = J_{orb}^2 - J_{orb}^2/\cos^2\iota + \varpi^2/4$$
$$= \varpi^2/4 - J_{orb}^2 \tan^2\iota = (1 - J_{orb}^2/J_o^2)\,\varpi^2/4 = (\tan^2\iota)\,\varpi^2/4.$$

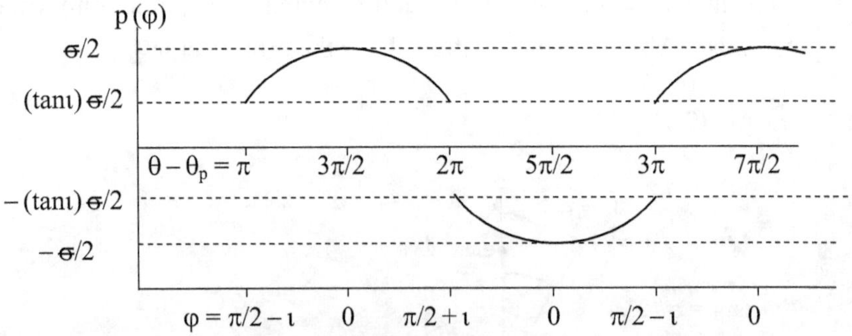

Figure A4.2. The Behavior Of $p(\varphi) = [r^4\varphi'^2 - r^2\varphi' \varpi(\cos\varphi)]^{1/2}$

The curve in Figure A4.2 is a sketch for $p(\varphi)$ as a function of θ for $\cos\varphi = -\sin\iota \cos(\theta - \theta_p)$. The value of $r^2\varphi'$ lies in the range set by $|r^2\varphi'| \le \varpi/2$ and expression 4-8, wherein $r^2\varphi'$ is non-zero, and the specific angular momentum component $p(\varphi) = [r^4\varphi'^{\,2} - r^2\varphi'\,\sigma\,(\cos\varphi)]^{1/2}$ abruptly changes its sign at $\cos\varphi = \pm\sin\iota$. However, both $r^4\varphi'^{\,2}$ and φ are continuous, even though φ' is discontinuous at the lower magnitude limits. This type of behavior is often found in the motion of bodies such as a symmetrical top wherein its spin axis rises and falls abruptly as precession ceases at a critical inclination angle.

Let us next use $r'\,r = r^2\theta'\,\varepsilon\,(\sin\theta)/[1 + \varepsilon\cos(\theta - \theta_p)]$ to estimate the effect of $\sigma\int r'\,r\cos\varphi\,d\varphi$ on $J_o^{\,2}$, where $r^2\theta' = J_{orb}/\sin^2\varphi$ and $d\varphi = -\sin\iota\sin\theta\,d\theta/\sin\varphi$ in equation 4-9 for $\cos\varphi = -\sin\iota\cos(\theta - \theta_p)$. The term $1/\sin^3\varphi$ is positive over the integral and has an average of $1/\sin^3\varphi_{av} \cong 1$ for small values of ι. Setting $1/[1 + \varepsilon\cos(\theta - \theta_p)] \cong 1 - \varepsilon\cos(\theta - \theta_p)$, we obtain the second form. Writing $\sin(\theta - \theta_p) = \sin\theta\cos\theta_p - \cos\theta\sin\theta_p$ and $\cos(\theta - \theta_p) = \cos\theta\cos\theta_p - \sin\theta\sin\theta_p$, the terms multiplied by 1 average to zero over the integral, as do all ε^2 terms except $\sin^2\theta\cos^2\theta$, whose integral provides a factor of $2\pi/8$ in the final form.

$$
\begin{aligned}
(4\text{-}9)\quad & \sigma\int r'\,r\cos\varphi\,d\varphi \\
&= \sigma\,J_{orb}\,\varepsilon\,(\sin^2\iota)\int[\sin(\theta - \theta_p)]\,(\sin\theta\cos\theta\,d\theta)/[(\sin^3\varphi)(1 + \varepsilon\cos(\theta - \theta_p))] \\
&\cong \sigma\,J_{orb}\,\varepsilon\,[(\sin^2\iota)/\sin^3\varphi_{av}]\,[\int[\sin(\theta - \theta_p)]\,[1 - \varepsilon\cos(\theta - \theta_p)]\,\sin\theta\cos\theta\,d\theta \\
&\cong \sigma\,J_{orb}\,\varepsilon^2\,(\sin^2\iota)\,(\sin^2\theta_p - \cos^2\theta_p)\,\pi/4.
\end{aligned}
$$

Division of equation 4-9 by $J_o^{\,2}$ provides the approximate magnitude of the perturbation relative to $J_o^{\,2}$. Its effect is greatest for the planet Mercury, whose unperturbed values are $J_o = 3\sigma/2$ and $\iota \cong 18.43$ degrees. Using the observed value of ε for Mercury of about 0.2056, we find the perturbation to be about three-tenths of one percent of the value of $J_o^{\,2}$ for $\theta_p = \pi/2$. We will ignore its effects on $J_o^{\,2}$, but recognize that it affects ι by a fraction of a degree. The effect appears to grow without bounds in time, but it is a single orbit term. Although we have used θ to model the integral, it is a function of φ, which is limited to $\pi/2 - \iota \le \varphi \le \pi/2 + \iota$. Moreover, the constant advancement of the perihelion of Mercury's orbit at a rate of about 1.5 degrees per century renders the perturbation oscillatory over a long period of time. It is zero at $\theta_p = \pi/4$, $3\pi/4$, $5\pi/4$, and $7\pi/4$. Between zeroes it increases smoothly in positive values, or decreases smoothly in negative values, before returning to zero. The period of the oscillation in $J_o^{\,2}$ is about 24,000 years.

APPENDIX 5 – ORBIT-LEVEL FRAME ROTATIONS

Let us consider a circular orbit of frequency W, planet mass μ_p, and orbital angular momentum $\mu_p\underline{\mathbf{J}}_{orb} = I_z W\,\underline{\mathbf{k}}_o = \mu_p r^2 W\,\underline{\mathbf{k}}_o$ aligned with the polar axis $\underline{\mathbf{k}}_o$ of the solar field. In an $\underline{\mathbf{ijk}}$ body frame inclined at an angle γ to $\underline{\mathbf{k}}_o$ in the solar $\underline{\mathbf{i}}_o\underline{\mathbf{k}}_o$ plane, $\mu_p\underline{\mathbf{J}}_{orb}$ appears as $\mu_p\underline{\mathbf{J}}_j + \mu_p\underline{\mathbf{J}}_k = -I_z W \sin\gamma\,\underline{\mathbf{j}} + I_z W\cos\gamma\,\underline{\mathbf{k}}$. But let us suppose that frame rotation of ξ' is occurring about the $\underline{\mathbf{k}}$ axis. As a result, $\mu_p\underline{\mathbf{J}}_k$ increases to $I_z\,(W\cos\gamma + \xi')\,\underline{\mathbf{k}}$, and the sum angular momentum vector in Figure A5.1 becomes $\mu_p\underline{\mathbf{J}}_{sum} = \mu_p\, r^2\,[\xi' \sin\gamma\,\underline{\mathbf{j}}_o + (W + \xi' \cos\gamma\,\underline{\mathbf{k}}_o)]$. The inclination δ of $\mu_p\underline{\mathbf{J}}_{sum}$ to $\underline{\mathbf{k}}_o$ is due to the composite of orbital motion and frame rotation, and will be slight so long as ξ' is small in comparison to W. A vector such as $\underline{\mathbf{J}}_{oz}$ expressed as a 3×1 column in the $\underline{\mathbf{i}}_o\underline{\mathbf{j}}_o\underline{\mathbf{k}}_o$ frame becomes $\underline{C}_{\gamma i}\,\underline{\mathbf{J}}_{oz}$ in the $\underline{\mathbf{ijk}}$ body frame, where $\underline{C}_{\gamma i}$ is specified by expression 5-1. The matrix $\underline{C}_{\xi k}$ further applies to $\underline{C}_{\gamma i}\,\underline{\mathbf{J}}_{oz}$ when frame rotation of ξ' occurs about $\underline{\mathbf{k}}$.

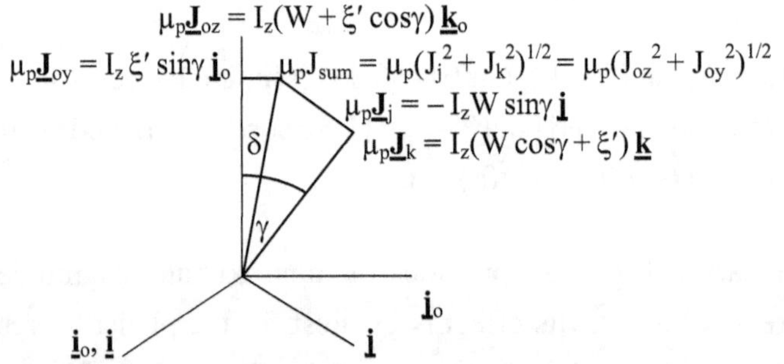

Figure A5.1. Angular Momentum Including Orbit-Level Frame Rotation

$$(5\text{-}1)\quad \underline{C}_{\gamma i} = \begin{bmatrix} 1 & 0 & 0 \\ 0 & \cos\gamma & -\sin\gamma \\ 0 & \sin\gamma & \cos\gamma \end{bmatrix}, \qquad \underline{C}_{\xi k} = \begin{bmatrix} \cos\xi & \sin\xi & 0 \\ -\sin\xi & \cos\xi & 0 \\ 0 & 0 & 1 \end{bmatrix}.$$

Figure 5.2 depicts the motion of the $\underline{\mathbf{ijk}}$ body frame. The $\underline{\mathbf{k}}$ axis remains in the $\underline{\mathbf{j}}_o\underline{\mathbf{k}}_o$ plane, and the $\underline{\mathbf{j}}$ axis lies initially in the same plane, inclined to $\underline{\mathbf{j}}_o$ at an angle $-\gamma$. However, the $\underline{\mathbf{i}}$ and $\underline{\mathbf{j}}$ axes are continually rotating about $\underline{\mathbf{k}}$ with a frequency of ξ'. The complementary elevation angle φ in the $\underline{\mathbf{i}}_o\underline{\mathbf{j}}_o\underline{\mathbf{k}}_o$ frame is provided by $\cos\varphi = -\sin\gamma \sin\theta$, where θ specifies the location of a planet in its orbit and φ fluctuates about $\pi/2$ over the orbit.

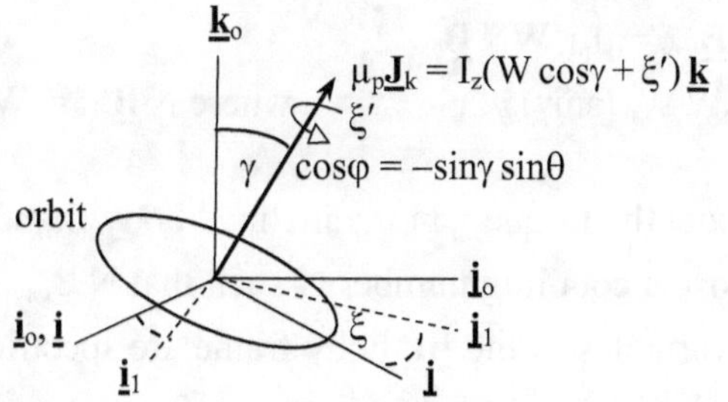

Figure A5.2. Inclined Body Frame Relationships

The Orbit-Level Torque Produced By $\underline{\boldsymbol{B}}_{gav}$

The frame rotation frequency ξ' is caused by a gravitational magnetic-like flux $\underline{\mathbf{B}}_g$, expression 5-2. Its orbital average $\underline{\mathbf{B}}_{gav}$ is provided by the second line of the expression, and the form of $\underline{\mathbf{B}}_{gav}$ is not changed when $\underline{\boldsymbol{C}}_{\gamma i}$ is applied. We will model the interaction of $\underline{\mathbf{B}}_{gav}$ with an orbiting mass μ_p similar to that of an electric current ϑ in a wire ring exposed to a magnetic flux $\underline{\mathbf{B}}$. The flux creates a torque $\underline{\boldsymbol{\tau}}$ on the inclined ring and causes it to turn in the direction of $\underline{\mathbf{B}}$, initially along the $-\underline{\mathbf{i}}_o$ axis. An accepted torque model is based on the magnetic moment $\underline{\boldsymbol{M}}_e = N\,\vartheta\,A_{ar}\,\underline{\mathbf{k}}$ for the ring, expression 5-3.

(5-2) $\underline{\mathbf{B}}_g = -\sigma\,(\cos\varphi)\,(-\sin\theta\,\underline{\mathbf{i}}_o + \cos\theta\,\underline{\mathbf{j}}_o)/(2r^2)$, where $\cos\varphi = -\sin\gamma\,\sin\theta$,
$\underline{\mathbf{B}}_{gav} = -\sigma\,(\sin\gamma)\,\underline{\mathbf{i}}_o/(4r^2)$, $\quad \underline{\boldsymbol{C}}_{\gamma i}\,\underline{\mathbf{B}}_{gav} = -\sigma\,(\sin\gamma)\,\underline{\mathbf{i}}/(4r^2)$,

(5-3) $\underline{\boldsymbol{M}}_e = N\,\vartheta\,A_{ar}\,\underline{\mathbf{k}}$, where $\underline{\boldsymbol{M}}_e$ is a magnetic moment, ϑ is the current in a loop, A_{ar} is the loop area, $\underline{\mathbf{k}}$ is perpendicular to the area, and N is a coupling number.

The current for an electron orbiting a nucleus is modeled classically as $\vartheta = e/t_p = eW/(2\pi)$, where e is the electron's charge, t_p is its orbital period, and W is the average orbital angular frequency. Thus, $\underline{\boldsymbol{M}}_e$ becomes $NeWr^2\,\underline{\mathbf{k}}/2$ for a circular orbit of radius r and area $A_{ar} = \pi r^2$, whose normal vector is $\underline{\mathbf{k}}$. The magnetic torque due to $\underline{\mathbf{B}}$ is properly modeled by a dyadic $\underline{\boldsymbol{M}}_e\,\underline{\mathbf{B}}$, which is the projection of $\underline{\boldsymbol{M}}_e$ onto $\underline{\mathbf{B}}$. Replacing e by μ_p, the equivalent gravitational moment $\underline{\boldsymbol{M}}_g$ takes the form of expression 5-4.

(5-4) $\underline{\boldsymbol{M}}_g = \mu_p\,NWr^2\,\underline{\mathbf{k}}/2$, for a magnetic-like gravitational moment,

$$(5\text{-}5)\quad \underline{\tau} = \boldsymbol{M}_g \underline{\mathbf{B}}_{gav} = \mu_p r^2 W N \underline{\mathbf{B}}_{gav}/2,$$
$$= -I_z W\, W_o\,(\sin\gamma)\,\underline{\mathbf{i}}/2, \qquad \text{where } N\,\underline{\mathbf{B}}_{gav} = W_o \sin\gamma\, \underline{\mathbf{i}}.$$

We will model the torque $\underline{\tau}$ in the inclined body frame by expression 5-5, where we propose a coupling number N such that $N\,\underline{\mathbf{B}}_{gav} = -W_o \sin\gamma\, \underline{\mathbf{i}}$. † The frame rotation variables in the $\underline{\mathbf{ijk}}$ body frame are specified by expression 5-6, where W in the $\underline{\mathbf{i}}_o\underline{\mathbf{j}}_o\underline{\mathbf{k}}_o$ frame appears as a frame rotation of ϕ' in the $\underline{\mathbf{ijk}}$ frame. A stable solution for Euler's equation 5-7 is provided by $\varpi_x = 0$, $\varpi_y = -W \sin\gamma$, and $\varpi_z = \xi' + W \cos\gamma$, with $\mu_p J_x = 0$, $\mu_p J_y = -(I_z/2)\, W \sin\gamma$, and $\mu_p J_z = I_z(W \cos\gamma + \xi')$, expression 5-8. The constancy of $\mu_p J_z = I_z W_o$ also determines the solution for the torque along $\underline{\mathbf{i}}$. Fluctuations of γ' may be very small, rather than precisely zero, with W varying slowly as a function of γ while keeping $\mu_p J_z = W \cos\gamma + \xi'$ constant. See Appendix 3, equation 3-11c.

$$(5\text{-}6)\quad \varpi_x = \gamma' = 0, \qquad \mu_p J_x = 0,$$
$$\varpi_y = -\phi' \sin\gamma, \qquad \mu_p J_y = -(I_z/2)\, W \sin\gamma, \quad \text{for } \phi' = W,$$
$$\varpi_z = \xi' + \phi' \cos\gamma, \qquad \mu_p J_z = I_z\,(\xi' + W \cos\gamma),$$

$$(5\text{-}7)\quad I_x\, \varpi_x' + \varpi_y\, \mu_p J_z - \varpi_z\, \mu_p J_y =$$
$$-(I_z/2)\,(W \sin\gamma)\,(\xi' + W \cos\gamma) = -I_z W\, W_o\,(\sin\gamma)/2,$$
$$I_y\, \varpi_y' + \varpi_z\, \mu_p J_x - \varpi_x\, \mu_p J_z = 0,$$
$$I_z\, \varpi_z' + \varpi_x\, \mu_p J_y - \varpi_y\, \mu_p J_x = d/dt\; I_z\,(W \cos\gamma + \xi') = 0,$$

$$(5\text{-}8)\quad I_z\,(W \cos\gamma + \xi') = I_z W_o = \mu_p J_{orb}, \text{ a constant,} \quad \text{and } \xi' = W_o - W \cos\gamma.$$

The torque in the body frame produces continuous rotation of the $\underline{\mathbf{i}}$ and $\underline{\mathbf{j}}$ axes about $\underline{\mathbf{k}}$. †† The $\underline{\boldsymbol{C}}_{\xi k}$ matrix may be applied to the angular momentum

---

† For the Earth, $I_z W\, W_o$ becomes $\mu_p\,(21\sigma^2/4)/r^2$ at γ_o. The coupling between $\underline{\mathbf{B}}_{gav}$ and the orbit reflects the occurrence of a constant current, and thus a type of resonance that occurs when $r^4 W^2$ takes the form $m\,\sigma^2/4$, where m is an integer. N may also be expressed as $(a_n/a_o)^{1/2}$, where $\sigma = (Mg_o a_o)^{1/2}$ and a_n is the average orbit radius.

†† Classical theory shows that a torque component of $I_z W^2 \sin\gamma \cos\gamma$ about the rotating $\underline{\mathbf{i}}$ axis is required to maintain the inclination γ of a dumbbell to the $\underline{\mathbf{k}}_o$ axis. See, *e.g.*, Becker, R. A., *Introduction to Theoretical Mechanics*, *supra*, at pp. 275, 292.

components and to the torque to obtain their forms in an $\underline{\mathbf{i}}_1\underline{\mathbf{j}}_1\underline{\mathbf{k}}_1$ inertial frame. Terms involving cosξ and sinξ appear in Euler's equation, but only those multiplied by cosξ survive cancellations for the $\underline{\mathbf{i}}_1$ component, and only those multiplied by sinξ survive for the $\underline{\mathbf{j}}_1$ component. The torque then becomes $I_z W \sigma (\sin\gamma)(\cos\xi\, \underline{\mathbf{i}}_1 - \sin\xi\, \underline{\mathbf{j}}_1)/(8r^2)$, and both components lead to the same result for ξ′. The term $I_z(W\cos\gamma + \xi') = I_z W_o = \mu_p J_{orb}$ is unaffected by $\underline{C}_{\xi k}$.

Our main objective, however, is to apply the inverse of $\underline{C}_{\gamma i}$ to the body frame, where $\xi'\,\underline{\mathbf{k}}$ in the $\underline{\mathbf{ijk}}$ frame becomes $\xi' \sin\gamma\, \underline{\mathbf{j}}_o + \xi' \cos\gamma\, \underline{\mathbf{k}}_o$ in the $\underline{\mathbf{i}}_o\underline{\mathbf{j}}_o\underline{\mathbf{k}}_o$ frame. The angular momentum sum $\mu_p \underline{\mathbf{J}}_{sum}$ in this frame is specified by expression 5-9. Figure A5.1 above displays these results, where the magnitude of $\mu_p \underline{\mathbf{J}}_{sum}$ can be expressed as $I_z |(W\,\underline{\mathbf{k}}_o + \xi'\,\underline{\mathbf{k}})|$ or as $|\mu_p \underline{\mathbf{J}}_{oy} + \mu_p \underline{\mathbf{J}}_{oz}|$ for the components $\mu_p \underline{\mathbf{J}}_{oz} = I_z (W + \xi' \cos\gamma)\, \underline{\mathbf{k}}_o$ and $\mu_p \underline{\mathbf{J}}_{oy} = I_z\, \xi' \sin\gamma\, \underline{\mathbf{j}}_o$ in the $\underline{\mathbf{i}}_o\underline{\mathbf{j}}_o\underline{\mathbf{k}}_o$ frame. The inclination of $\underline{\mathbf{J}}_{sum}$ to $\underline{\mathbf{k}}_o$ is $\delta = \text{arc-tan}[(\xi' \sin\gamma)/(W + \xi' \cos\gamma)]$.

(5-9) $\mu_p \underline{\mathbf{J}}_{sum} = \mu_p r^2 [\xi' \sin\gamma\, \underline{\mathbf{j}}_o + (W + \xi' \cos\gamma\, \underline{\mathbf{k}}_o)]$.

The inertial values of $I_z W_o = \mu_p (21/4)^{1/2} \sigma$ and $I_z W_{oz} = \mu_p (24/4)^{1/2} \sigma$ for the Earth provided by our wave theory, together with $I_z W_o = \mu_p (32/4)^{1/2} \sigma$ and $I_z W_{oz} = \mu_p (35/4)^{1/2} \sigma$ for Mars, allow us to compute γ, ξ′/W, and the components of $\underline{\mathbf{J}}_{sum}$ for these two orbits. Since the inertial values remain constant for all values of γ, we would find γ_{oz} to be $\text{arc-cos}(21/24)^{1/2} \cong 20.71$ degrees for the Earth if δ, and thus ξ′, were to become zero. This end can be achieved by allowing the average angular frequency W to vary with γ so that ξ′ is aliased into W, which becomes W_{oz} at $\gamma \cong 20.706$ degrees for the Earth. Since there is no torque along the $\underline{\mathbf{k}}_o$ axis, the component $I_z (W + \xi' \cos\gamma)$ is another constant $I_z W_{oz}$, equation 5-10. Multiplying equation 5-8 by cosγ and subtracting it from 5-10, we obtain equation 5-11 for W as a function of γ.

(5-10) $W + \xi' \cos\gamma = W_{oz}$,

(5-11) $W = W_o (W_{oz}/W_o - \cos\gamma)/\sin^2\gamma$.

Inputting $W_{oz}/W_o = (24/21)^{1/2} = 1.069$, we verify for the Earth that ξ'_{oz} is zero at $\gamma_{oz} \cong 20.706$ degrees. We also find that $W = W_o$ and $\xi'_o \cong 0.0746\, W_o$ at $\gamma_o \cong 22.27$ degrees, so that $\xi'_o \cos\gamma_o \cong 0.069\, W_o$. The angle δ is about 1.51

degrees. The corresponding values for Mars are $W_{oz}/W_o = (35/32)^{1/2} \cong 1.046$ and $\xi'_{oz} = 0$ at $\gamma_{oz} \cong 17.02$ degrees. The Martian values at $W = W_o$ are $\xi'_o \cong 0.0482\, W_o$ at $\gamma_o \cong 17.88$ degrees, with the angle $\delta \cong 0.77$ degrees.

Although $\underline{\mathbf{B}}_{gav}$ creates continuous frame rotation of ξ' about $\underline{\mathbf{k}}$ in the $\underline{\mathbf{ijk}}$ frame, the rotation is offset by a non-orbital angular momentum component in the $\underline{\mathbf{i}}_o\underline{\mathbf{j}}_o\underline{\mathbf{k}}_o$ frame, and the orbit remains stable at the inclination γ_{oz}.

The Orbit-Level Lagrangian And The Potential Q

(5-12) $$L_{sum}/\mu_p = [r' + (\sigma/2)(\sin\varphi)/r]^2/2 + [r\,\varphi' + (\sigma/2)(\cos\varphi)/r]^2/2 + Mg_o/r + [r\,\theta' + a_n^2(\xi'\cos\gamma)/r]^2(\sin^2\varphi)/2 + a_n^4\,\xi'^2(\sin^2\gamma)/(2r^2) - Q,$$

We offer expression 5-12 as the orbit-level Lagrangian L_{sum} that includes the effects of the vector potential $\underline{\mathbf{A}}_g$ and frame rotation ξ', and reduces to the accepted form when they are not present. The nominal potential energy is $V = -\mu_p Mg_o/r$, and $\mu_p Q$ is a potential form whose partial derivative with respect to γ models the torque that produces ξ'. First taking the partial of L_{sum} with respect to ξ', where none of the terms depend on ξ, we obtain equation 5-13. Evaluating the integral at the average of $\varphi = \pi/2$ for a circular orbit, equation 5-14 is the solution, where $\mu_p a_n^2\,\theta' = I_z W$ and $\mu_p a_n^2\,\xi' = I_z\,\xi'$ are the orbit-averaged values. The integration constant is $\mu_p J_{orb} = I_z W_o$.

(5-13) $$\mu_p\, d/dt\,[r\,\theta' + a_n^2(\xi'\cos\gamma)/r]\,(\sin^2\varphi)\,a_n^2(\cos\gamma)/r + a_n^4\,\xi'(\sin^2\gamma)/r^2] = 0,$$

(5-14) $$I_z(W\cos\gamma + \xi') = \text{constant} = I_z W_o = \mu_p J_{orb},$$

(5-15) $$\mu_p[r\,\theta' + a_n^2(\xi'\cos\gamma)/r]\,[a_n^2(-\xi'\sin\gamma)/r]\,(\sin^2\varphi) + \mu_p\,a_n^4\,\xi'^2(\sin\gamma)(\cos\gamma)/r^2 - \mu_p\,\partial Q/\partial\gamma = 0,$$

(5-16) $$I_z W\,\xi'\sin\gamma = -\mu_p\,\partial Q/\partial\gamma.$$

Next taking the partial of L_{sum} with respect to γ and ignoring γ'' at the orbit level, we obtain equation 5-15. Using orbital averages, expression 5-16 specifies ξ' as a function of Q. Multiplying both sides of equation 5-14 by $W\sin\gamma$, we can express $W\,\xi'\sin\gamma$ as a function of W, W_o, and γ. Equating the result to $-\mu_p\partial Q/\partial\gamma$, we obtain expression 5-17 for $\mu_p Q$, where the partial derivative allows us to express I_z as $\mu_p r^2$. Thus, $r^2 W_o^{\,2}/2$ depends only on r, and $\mu_p Q$ becomes $-I_z\xi'^{\,2}/2$ for a circular orbit when r is averaged to a_n. Both $\mu_p Q$ and ξ' become zero at γ_{oz}, where $W_{oz} = W_o/\cos\gamma_{oz}$ and $J_{orb} = J_{oz}\cos\gamma_{oz}$.

(5-17) $\mu_p Q = -\,\mu_p r^2 (W_o - W\cos\gamma)^2/2.$

Taking the partial of L_{sum} with respect to θ', we obtain equation 5-18 with equation 5-19 for its solution. We may identify the orbital average of $\mu_p r^2\theta' \sin^2\varphi$ as $I_z W$, and $\mu_p a_n^{\,2}\,\xi' \cos\gamma \sin^2\varphi$ as $I_z\,\xi' \cos\gamma$, which sum to the constant $\mu_p J_{oz}$ as W and ξ' are allowed to vary.

(5-18) $\mu_p\, d/dt\,[(r^2\theta' + a_n^{\,2}\,\xi' \cos\gamma)\sin^2\varphi] = \partial L_{sum}/\partial\theta = 0,$

(5-19) $\mu_p\,(r^2\theta' + a_n^{\,2}\,\xi' \cos\gamma)\sin^2\varphi = \text{constant} = I_z\,(W + \xi' \cos\gamma) = \mu_p\, J_{oz}.$

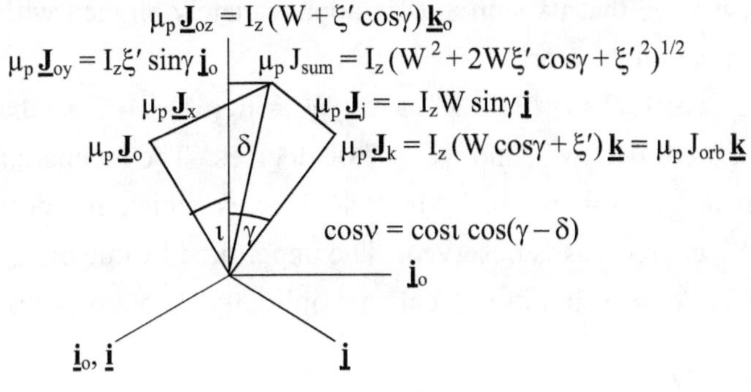

Figure A5.3. Angular Momentum Configuration For The Earth And Mars

The partials of L_{sum} with respect to φ' and φ provide equation 5-20, with $r^2\varphi'$ varying to maintain $r^2\varphi'\,[r^2\varphi' - \sigma\,(\cos\varphi)/2] + J_{orb}^{\,2}/\sin^2\varphi - \sigma^2/4 = J_{orb}^{\,2}$. Equation 5-21 is its perturbed solution. See Appendix 4, discussing the effects of the vector potential $\underline{\mathbf{A}}_g$ on the end orbit vector $\mu_p\,\underline{\mathbf{J}}_o$ depicted in Figure A5.3.

(5-20) $[1/(2r^3\varphi')]\; d/dt\,(r^4\varphi'^{\,2} + J_{oz}^{\,2}/\sin^2\varphi) = \sigma\, r'\,(\cos\varphi)/(2r^2),$

(5-21) $r^4\varphi'^{\,2} + (r^2\theta' + a_n^{\,2}\,\xi' \cos\gamma)^2 \sin^2\varphi - \sigma\!\int r'\, r\cos\varphi\, d\varphi = J_o^{\,2}.$

Lastly taking the partials of L_{sum} with respect to r′ and r produces equation 5-22, where the cross terms $r\,\varphi'\,[(ϭ/2)\,(\cos\varphi)/r]$ and $r\,\theta'\,[a_n^{\,2}\,(\xi'\cos\gamma)/r]$ are independent of r. At $\varphi = \pi/2$, the terms $a_n^{\,4}\,\xi'^{\,2}/r^3$ and $\partial Q/\partial r$ in the second line sum to zero, and we are left with the expression of $J_{orb}^{\;2}$ provided by equation 4-6 in Appendix 4, where $r^4\varphi'^{\,2} - ϭ\,r^2\varphi'\,(\cos\varphi)/2 - ϭ^2/4$ is also zero.

$$
\begin{aligned}
(5\text{-}22)\quad d^2r/dt^2 &= [r^4\varphi'^{\,2} + r^4\theta'^{\,2}\sin^2\varphi - ϭ\,r^2\varphi'\,(\cos\varphi)/2 - ϭ^2/4]/r^3 - Mg_o/r^2 \\
&\quad + [-a_n^{\,4}\,\xi'^{\,2}\,(\cos^2\gamma)\,(\sin^2\varphi)/r^3 - a_n^{\,4}\,\xi'^{\,2}\,(\sin^2\gamma)/r^3 - \partial Q/\partial r] \\
&= J_{orb}^{\;2}/r^3 - Mg_o/r^2, \qquad \text{where } J_{orb}^{\;2} = r^4\theta'^{\,2} = a_n^{\,4}\,W^2 \text{ at } \varphi = \pi/2.
\end{aligned}
$$

Using our wave theory results to express $J_o = 5ϭ/2$ with $J_{oz} = (24)^{1/2}\,ϭ/2$ and $J_x^{\;2} = J_o^{\;2} - J_{oz}^{\;2} = ϭ^2/4$ for the Earth, we obtain $\iota \cong 11.31$ degrees from expression 4-7 in Appendix 4. This is the Earth's stabilized orbit configuration, with the angle between $\underline{\mathbf{J}}_o$ and $J_{orb}\,\underline{\mathbf{k}}$ becoming ν_{oz} = arc-cos($\cos\iota\,\cos\gamma_{oz}$) $\cong$ 23.47 degrees. Since the Earth's spin obliquity is about 23.45 degrees at the present time, we see that its spin axis is approximately aligned with its orbital angular momentum vector.

For Mars, we have $J_{oz}/J_{orb} = (35/32)^{1/2}$ and $I_zW_o = \mu_p\,(32)^{1/2}\,ϭ/2$, so that $\gamma_o \cong 17.88$ degrees, $\xi'_o \cong 0.0482\,W_o$, and $\gamma_{oz} \cong 17.02$ degrees. The inclination of Mars' orbital axis to $\underline{\mathbf{k}}_o$ is ι = arc-tan(1/6) $\cong$ 9.46 degrees, which differs from the Earth's by 1.85 degrees, as is observed. The unperturbed value of ν_{oz} for Mars is about 19.41 degrees, but its present spin obliquity is 25.2 degrees.

Summary Of The Effects Of $\underline{\mathbf{A}}_g$ And $\underline{\mathbf{B}}_g$

In a spin-based $\underline{\mathbf{ijk}}$ frame, where the orbital angular momentum vector $I_zW\,\underline{\mathbf{k}}$ is inclined to $\underline{\mathbf{k}}_o$ at an angle γ in the $\underline{\mathbf{j}}_o\underline{\mathbf{k}}_o$ plane, frame rotation of ξ' occurs due to the torque created by $\underline{\mathbf{B}}_{gav}$. The result adds to $I_zW\,\underline{\mathbf{k}}$ to produce a sum vector of $\mu_p\,\underline{\mathbf{J}}_{sum} = I_z[\xi'\sin\gamma\,\underline{\mathbf{j}}_o + (W + \xi'\cos\gamma)\,\underline{\mathbf{k}}_o]$ for the Earth and Mars. The Lagrangian term $a_n^{\,2}\xi'\cos\gamma\,\underline{\mathbf{k}}_o/r$ couples with $r\,\theta'\,\underline{\mathbf{k}}_o$ in the $\underline{\mathbf{i}}_o\underline{\mathbf{j}}_o\underline{\mathbf{k}}_o$ frame as the term $a_n^{\,2}\xi'\,(\sin\gamma)\,\underline{\mathbf{j}}_o/r$ stands alone. Their squares appear in the orbital energy, expression 5-23. In the inertial frame $\underline{\mathbf{J}}_{sum}$ becomes $\underline{\mathbf{J}}_{oz}$ and aligns with $\underline{\mathbf{k}}_o$ when $I_zW\,\underline{\mathbf{k}}$ is inclined to $\underline{\mathbf{k}}_o$ at the angle γ_{oz}, and the frame rotation is aliased into $\underline{\mathbf{J}}_{oz}$. The component $I_z(W\cos\gamma + \xi')\,\underline{\mathbf{k}}$ in the body frame remains constant in magnitude at $I_zW_o = \mu_p\,J_{orb}$ as γ varies. The vector potential $\underline{\mathbf{A}}_g$ couples with the real velocity components $\underline{\mathbf{v}}_r$ and $\underline{\mathbf{v}}_\varphi$, and adds a factor of $\varsigma^2/4$ to the formulation of orbital kinetic energy in expression 5-23. As a result, the square magnitude of the orbit vector is $\underline{\mathbf{J}}_o^{\,2} = \underline{\mathbf{J}}_{oz}^{\,2} + \varsigma^2/4$ in inertial space, and it is inclined to $\underline{\mathbf{k}}_o$ at an angle ι in the $\underline{\mathbf{i}}_o\underline{\mathbf{k}}_o$ plane.

$$(5\text{-}23)\quad E_{sum}/\mu_p = [r' + (\varsigma/2)(\sin\varphi)/r]^2/2 + [r\,\varphi' + (\varsigma/2)(\cos\varphi)/r]^2/2$$
$$+ (\sin^2\varphi)\,[r\,\theta' + a_n^{\,2}(\xi'\cos\gamma)/r]^2/2 + a_n^{\,4}\,\xi'^{\,2}(\sin^2\gamma)/(2r^2)$$
$$- r^2(W_o - W\cos\gamma)^2/2 - Mg_o/r.$$

Neither the term $\mu_p\varsigma^2/4$ produced by $\underline{\mathbf{A}}_g$ nor $I_z\xi'\,\underline{\mathbf{k}}$ is orbital angular momentum, even though the orbit vector lies in the direction of $\underline{\mathbf{J}}_o$. The term $2r^2\theta'\,a_n^{\,2}\xi'\,(\cos\gamma)/r^2$ due to the coupling of θ′ and ξ′ vanishes when we take the partial of L_{sum} with respect to r, and we are left with the orbital term $J_{orb}^{\;2} = r^2\varphi'\,[r^2\varphi' - \varsigma(\cos\varphi)/2] + J_{orb}^{\;2}/\sin^2\varphi - \varsigma^2/4$, as $\partial Q/\partial r$ and $a_n^{\,4}\,\xi'^{\,2}/r^2$ cancel in the r-equation. The term $\varsigma\, r'\, r\,\varphi'\cos\varphi$ is a small perturbation for nominal orbit inclinations of ι to the field axis $\underline{\mathbf{k}}_o$ and modest eccentricities. Thus, the solution for r is essentially unchanged from the classical result, but happily, the orbit inclinations can no longer be regarded as random values.

Modeling The Ecliptic In The Solar Field

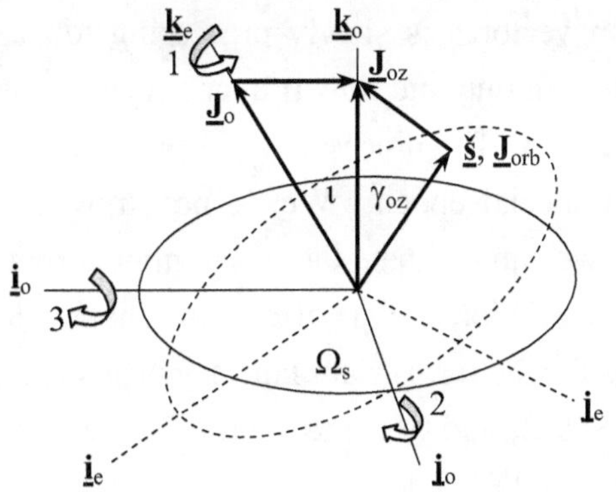

Figure A5.4. The Ecliptic Frame When $\underline{\check{\mathbf{s}}}$ And $\underline{\mathbf{J}}_{orb}$ Are Aligned

In order to validate our inclination predictions in Chapter 6 of the text, we must relate the inertial $\underline{\mathbf{i}}_o\underline{\mathbf{j}}_o\underline{\mathbf{k}}_o$ frame to the $\underline{\mathbf{i}}_e\underline{\mathbf{j}}_e\underline{\mathbf{k}}_e$ ecliptic frame. The $\underline{\mathbf{i}}_e\underline{\mathbf{j}}_e\underline{\mathbf{k}}_e$ frame has been formed by aligning its $\underline{\mathbf{k}}_e$ axis with the Earth's orbit vector

$\underline{\mathbf{J}}_o$, and setting the $\underline{\mathbf{i}}_e$ axis by crossing the spin vector $\underline{\check{\mathbf{s}}}$ into $\underline{\mathbf{J}}_o$. Since the spin obliquity is about 23.45 degrees at the present time, $\underline{\check{\mathbf{s}}}$ appears in vector form as $\check{s}\,(\sin 23.45^\circ\ \underline{\mathbf{j}}_e + \cos 23.45^\circ\ \underline{\mathbf{k}}_e)$ in the ecliptic frame. When $\underline{\check{\mathbf{s}}}$ and $\underline{\mathbf{J}}_{orb}$ are aligned, the frame relationships are depicted in Figure A5.4, where the inertial $\underline{\mathbf{i}}_o\underline{\mathbf{j}}_o$ plane intersects the ecliptic $\underline{\mathbf{i}}_e\underline{\mathbf{j}}_e$ plane along a node longitude of Ω_s.

If $\underline{\check{\mathbf{s}}}$ were lying in the $\underline{\mathbf{j}}_o\underline{\mathbf{k}}_o$ plane at the present time, we could reach the $\underline{\mathbf{i}}_o\underline{\mathbf{j}}_o\underline{\mathbf{k}}_o$ frame from the ecliptic by straightforward frame rotations. We would first perform a backward rotation of Ω_s about $\underline{\mathbf{k}}_e$ to align the $\underline{\mathbf{j}}_e$ axis with $\underline{\mathbf{i}}_o$, as indicated by the number 1 in Figure A5.4. Next rotating about $\underline{\mathbf{j}}_o$ by $\iota = -11.31$ degrees, indicated by 2 in the figure, the spin vector would take the form $\underline{\check{\mathbf{s}}} = \check{s}\,(\sin\gamma\ \underline{\mathbf{j}}_o + \cos\gamma\ \underline{\mathbf{k}}_o) = \underline{U}_2\underline{U}_1\ \underline{\check{\mathbf{s}}}_o$ in the $\underline{\mathbf{i}}_o\underline{\mathbf{j}}_o\underline{\mathbf{k}}_o$ frame, where $\underline{U}_2$ and $\underline{U}_1$ are 3×3 matrices provided by expression 5-24 for position vectors expressed as 3×1 columns. The requirement for the $\underline{\mathbf{i}}_o$ component of $\underline{\check{\mathbf{s}}}$ to be zero in the solar frame provides $\Omega_s \cong 62.57$ degrees. If γ were not γ_{oz} at the present epoch, a third rotation $\underline{U}_3$ in the figure based on $\gamma' t$ would align $\underline{\check{\mathbf{s}}}$ with $J_{orb}\,\underline{\mathbf{k}}$.

$$(5\text{-}24)\quad \underline{U}_3\,\underline{U}_2\,\underline{U}_1 = \begin{bmatrix} 1 & 0 & 0 \\ 0 & \cos\gamma' t & \sin\gamma' t \\ 0 & -\sin\gamma' t & \cos\gamma' t \end{bmatrix} \begin{bmatrix} \cos\iota & 0 & \sin\iota \\ 0 & 1 & 0 \\ -\sin\iota & 0 & \cos\iota \end{bmatrix} \begin{bmatrix} \sin\Omega_s & -\cos\Omega_s & 0 \\ \cos\Omega_s & \sin\Omega_s & 0 \\ 0 & 0 & 1 \end{bmatrix}.$$

Unhappily, the ecliptic is not an inertial reference frame since the Earth's spin vector $\underline{\check{\mathbf{s}}}$ is slowly precessing to the west and simultaneously nutating (nodding up and down) over very long periods of time, comparable to a spinning top. The precession of $\underline{\check{\mathbf{s}}}$ prevents it from lying in the $\underline{\mathbf{j}}_o\underline{\mathbf{k}}_o$ plane except at a special epoch. We do not know a priori when the alignment occurs, but we can estimate the misalignment by fitting the planetary data to the theory predictions. Our efforts in Chapter 6 of the text show that the inclinations for the terrestrial planetary orbits are set by $\tan\iota = (1/2)/(j + 1/2)$ for j = 1, 3/2, 2, and 5/2, in order. The respective values of ι are 18.43, 14.04, 11.31, and 9.46 degrees. We are thus able to obtain a good data fit as provided in Table 3 of Chapter 6 by locating the intersection of the $\underline{\mathbf{i}}_o\underline{\mathbf{j}}_o$ plane with the ecliptic $\underline{\mathbf{i}}_e\underline{\mathbf{j}}_e$ plane along a line running from approximately $\Omega_{s-} = 42.0$ degrees to $\Omega_{s+} = 222.0$ degrees. We therefore conclude that the Earth's spin vector has precessed by approximately 62.6 – 42.0 = 20.6 degrees since the time that it last lay in the $\underline{\mathbf{j}}_o\underline{\mathbf{k}}_o$ inertial plane of the solar field.

APPENDIX 6 – STANDING GRAVITY WAVE SOLUTIONS

We shall use energy expression 5-19 in Appendix 5 to obtain standing wave solutions in the solar gravitational field. We first form the vector ɵ$\underline{S}/r$ = ɵ$(\underline{s}_A - \underline{s})/r$, expression 6-1, where $\underline{s}_A$ and $\underline{s}$ are pure number vectors. The vector ɵ$\underline{s}/r$ models a component of frame rotational angular momentum, and ɵ $\underline{s}_A/r$ models the vector potential $\underline{\mathbf{A}}_g$. Both vectors couple with velocity components. Figure A6.1 depicts the specific angular momentum vectors in the $\underline{\mathbf{i}}_o\underline{\mathbf{j}}_o\underline{\mathbf{k}}_o$ frame for a planet of mass μ_p that is undergoing frame rotation. The vector $\underline{\mathbf{J}}_{sum}$ is inclined in the $\underline{\mathbf{j}}_o\underline{\mathbf{k}}_o$ plane at an angle δ to $\underline{\mathbf{k}}_o$, and is the sum of the body vectors $\underline{\mathbf{J}}_j$ and $\underline{\mathbf{J}}_k$. The vector $\underline{\mathbf{J}}_{oz}$ is the projection of $\underline{\mathbf{J}}_{sum}$ on the $\underline{\mathbf{k}}_o$ axis. $\underline{\mathbf{A}}_g$ is represented by $\underline{\mathbf{J}}_x$ lying on the $\underline{\mathbf{i}}_o$ axis, which is orthogonal to $\underline{\mathbf{J}}_{sum}$ and $\underline{\mathbf{J}}_{oz}$, and its square subtracts from $\underline{\mathbf{J}}_o^2$ to produce $\underline{\mathbf{J}}_{oz}$. We emphasize that we are now modeling <u>angular momentum vectors as functions of r, φ, and θ</u>, and not the planetary positions. Thus, the statement in our analysis that "φ approaches zero" means that the total angular momentum vector $\mu_p\underline{\mathbf{J}}_{sum}$ aligns with the $\underline{\mathbf{k}}_o$ axis at $\delta = \varphi = 0$.

(6-1) ɵ$\underline{S}/r$ = ɵ$(\underline{s}_A - \underline{s})/r$ = ɵ$(s_A \cos\theta\, \underline{\mathbf{i}}_o + s_A \sin\theta\, \underline{\mathbf{j}}_o - s\, \underline{\mathbf{k}}_o)/r$
= ɵ$[(s_A \sin\varphi - s \cos\varphi)\, \underline{\hat{\mathbf{u}}}_r + (s_A \cos\varphi + s \sin\varphi)\, \underline{\hat{\mathbf{u}}}_\varphi]/r$.

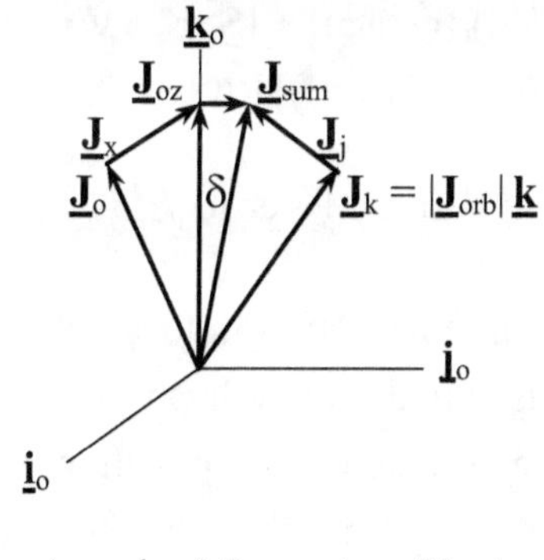

Figure A6.1. Angular Momentum Vector Relationships

The Time-Independent Equation

We showed above that taking the partial derivative of a state function twice with respect to a spatial coordinate produces the negative of that velocity component squared, with velocity and position acting as Fourier complements. We will include the velocity-like vector $i \sigma \underline{\mathbf{S}}/r$ in the wave equation as an addition to $-i\sigma\nabla$, with the square of the sum being negative. We also argued that these types of relationships continue to hold when terms such as $V = -\mu_p M g_o/r$ are included in the wave equation. Including the term $\mu_p Q = -\mu_p r^2 (W_o - W \cos\gamma)^2/2$ in the expression of energy, let us specify the *Hamiltonian operator* by the left side of equation 6-2. Equation 6-3 results when we expand and allow ∇ to operate on sequential terms. In equation 6-4 the state function Ψ is expressed as the product $R(r)\, \Phi(\varphi)\, \Theta(\theta)$ of separable functions of r, φ, and θ, and we will express E as E_n for discrete energy states. Treating $\Phi(\varphi)$ and $\Theta(\theta)$ as functions of φ and θ only, we may replace $\partial\Phi/\partial\varphi$ and $\partial\Theta/\partial\theta$ by $d\Phi/d\varphi$ and $d\Theta/d\theta$.

$$(6\text{-}2) \quad [-\mu_p \sigma^2 (\nabla + \underline{\mathbf{S}}/r) \bullet (\nabla + \underline{\mathbf{S}}/r)/2 - \mu_p M g_o/r + \mu_p Q]\, \Psi = E_n \Psi,$$

$$(6\text{-}3) \quad \nabla^2\Psi + 2\,(\underline{\mathbf{S}}/r) \bullet \nabla\Psi + \Psi\, \nabla \bullet (\underline{\mathbf{S}}/r) + [\underline{\mathbf{S}}^2/r^2 + (2/\sigma^2)\,(M g_o/r - Q + E_n/\mu_p)]\, \Psi = 0,$$

$$\begin{aligned}(6\text{-}4) \quad & [\Phi\,\Theta/r^2]\, [\partial/\partial r\, (r^2\, \partial R/\partial r)] + [R\,\Phi/(r^2 \sin^2\varphi)]\, [d^2\Theta/d\theta^2] \\ & + [R\,\Theta/(r^2 \sin\varphi)]\, [d/d\varphi\, (\sin\varphi\; d\Phi/d\varphi)] \\ & + [2/r]\, [\Phi\, \Theta\, S_r\, \partial R/\partial r + (R\, \Theta\, S_\varphi\, d\Phi/d\varphi)/r \\ & + (R\, \Phi\, S_\theta\, d\Theta/d\theta)/(r \sin\varphi)] + [\underline{\mathbf{S}}^2/r^2 + \nabla \bullet (\underline{\mathbf{S}}/r) \\ & + (2/\sigma^2)\, (M g_o/r - Q + E_n/\mu_p)]\, R\Phi\Theta = 0,\end{aligned}$$

$$(6\text{-}5) \quad \nabla \bullet \underline{\mathbf{S}}/r = [s_A\, (1/\sin\varphi - \sin\varphi) + s \cos\varphi]/r^2.$$

The divergence of $\underline{\mathbf{S}}/r$ in equation 6-4, does not vanish unless $s_A = s = 0$. Its form, expression 6-5, is singular at both $\varphi = 0$ and $r = 0$ for $s_A \neq 0$, and at $r = 0$ when $s \neq 0$. In electromagnetic applications the divergence of $\underline{\mathbf{A}}$ is often set to zero, but that option is not available here.

The Azimuth Angle Solution Of The Wave Equation

The term $S_\theta/\sin\varphi = (-s_x \sin\theta + s_y \cos\theta)/\sin\varphi$ would seem to prevent the separation of $\Phi(\varphi)$ from $\Theta(\theta)$ in equation 6-4. But since $s_x = s_A \cos\theta$ and s_y

$= s_A \sin\theta$, it follows that $S_\theta \equiv 0$, and multiplication of equation 6-4 by the factor $(r^2 \sin^2\varphi)/(R\Phi\Theta)$ isolates the only θ-dependent term $(1/\Theta)\, d^2\Theta/d\theta^2$. In order to be independent of r and φ, this term must be a constant called the *azimuth separation constant.* Equation 6-6 results when we set the constant to $-m^2$. The negative sign results in oscillatory behavior of Θ in equation 6-7, rather than exponential decay. If we require that $\Theta(\theta + 2\pi) = \Theta(\theta)$, then m should be an integer. But consistent with modeling half-integral values of σ and the appearances of the squares of θ and φ terms as observables, we require only that $\Theta^2(\theta)$ be unambiguous, so that m may be a half-integer.

(6-6) $\quad d^2\Theta/d\theta^2 + m^2\, \Theta = 0,$ where m is a half-integer or integer,

(6-7) $\quad \Theta(\theta) = C_1 \sin(m\theta) + C_2 \cos(m\theta),$ where C_1 and C_2 are constants.

The Elevation Angle Solution

Upon Multiplying equation 6-4 by $r^2/(R\Phi\Theta)$, replacing $(1/\Theta)\, d^2\Theta/d\theta^2$ by $-m^2$, setting $\underline{\mathbf{S}}^2$ to $s_A{}^2 + s^2$, and inputting $\nabla \cdot \mathbf{S}/r$, we obtain equation 6-8. Multiplying the equation by $R(r)/r^2$, equation 6-8 becomes 6-9, where the coefficient of $1/r^2$ is an angular momentum type of function f(φ), equation 6-10. The term Q is a function of r and the angle γ, and only implicitly of φ.

(6-8)
$$\begin{aligned}&(1/R)\, \partial/\partial r\, (r^2\, \partial R/\partial r) + [1/(\Phi \sin\varphi)]\, [d/d\varphi\, (\sin\varphi\, d\Phi/d\varphi)]\\ &\quad - m^2/\sin^2\varphi + 2r\, (s_A \sin\varphi - s \cos\varphi)\, (1/R)\, \partial R/\partial r\\ &\quad + 2\, (s_A \cos\varphi + s \sin\varphi)\, (1/\Phi)\, d\Phi/d\varphi] + s_A\, (1/\sin\varphi - \sin\varphi)\\ &\qquad + s_A{}^2 + s^2 + s \cos\varphi + (2\, r^2/\sigma^2)\, (Mg_o/r - Q + E_n/\mu_p) = 0,\end{aligned}$$

(6-9)
$$\begin{aligned}&(1/r^2)\, \partial/\partial r\, (r^2\, \partial R/\partial r) + (2/r)\, (s_A \sin\varphi - s \cos\varphi)\, \partial R/\partial r\\ &\qquad + [f(\varphi)/r^2 + 2\, (Mg_o/r - Q + E_n/\mu_p)/\sigma^2]\, R = 0,\end{aligned}$$

(6-10)
$$\begin{aligned}f(\varphi) = {}& 1/(\Phi \sin\varphi)]\, [d/d\varphi\, (\sin\varphi\, d\Phi/d\varphi)] - m^2/\sin^2\varphi\\ &+ 2\, (s_A \cos\varphi + s \sin\varphi)\, (1/\Phi)\, d\Phi/d\varphi\\ &+ s_A\, (1/\sin\varphi - \sin\varphi) + s_A{}^2 + s^2 + s \cos\varphi.\end{aligned}$$

The factor $s_A \sin\varphi - s \cos\varphi$ in equation 6-9 can be combined with f(φ), and both can be put in a single location by substituting $R(r) = r^b\, R_1(r)$, where

$b = -s_A \sin\varphi + s \cos\varphi$.[†] Equation 6-11 results, and $R_1(r)$ will be a function of r only with $\partial R_1/\partial r$ being replaced by dR_1/dr whenever the term $l\,(l+1)\,\not\sigma^2$ defined by expression 6-12 is a constant. Inputting $f(\varphi)$ from expression 6-10 to equation 6-12 and rearranging somewhat, we obtain equation 6-13.

$$\text{(6-11)}\quad (1/r^2)\,\partial/\partial r\,(r^2\,dR_1/dr) + [-l\,(l+1)/r^2 + 2(Mg_o/r - Q + E_n/\mu_p)/\not\sigma^2]\,R_1 = 0,$$

$$\text{(6-12)}\quad l\,(l+1) = -f(\varphi) + (s_A \sin\varphi - s \cos\varphi)^2 + (s_A \sin\varphi - s \cos\varphi),$$

$$\begin{aligned}\text{(6-13)}\quad & [1/(\Phi \sin\varphi)]\,[d/d\varphi\,(\sin\varphi\,d\Phi/d\varphi)] - m^2/\sin^2\varphi \\ & + 2\,(s_A \cos\varphi + s \sin\varphi)\,(1/\Phi)\,d\Phi/d\varphi \\ & + s_A\,(1/\sin\varphi - \sin\varphi) + s \cos\varphi + s_A{}^2 + s^2 \\ & - (s_A \sin\varphi - s \cos\varphi)^2 - (s_A \sin\varphi - s \cos\varphi) = -l\,(l+1).\end{aligned}$$

First considering the case when s_A and s are zero, equation 6-13 reduces to 6-14. The solutions for Φ are the associated Legendre functions of degree l and order m. Two independent solutions exist, but only the first kind, based on the mth derivative of polynomials of degree l in powers of $\cos\varphi$ and expressed as $\Phi(\varphi) = P^m{}_l(\cos\varphi)$, are regular at both $\varphi = 0$ and $\varphi = \pi$. The solutions require l and m to be integers with $m \le l$, indicating a form of degeneracy. However, the solution $\Phi(\varphi) = \sin^p\varphi \cos^q\varphi$ is also available, where p and q may not be integers. Inputting this form to equation 6-14, we obtain 6-15. Regularity on the equation's right side as φ approaches zero requires p^2 to be m^2 for $m \neq 0$, and further requires q to be 0 or 1 for non zero values of φ. Expression 6-16 is the solution for $\Phi(\varphi)$ when q is zero.

$$\begin{aligned}\text{(6-14)}\quad & (1/\sin\varphi)\,d/d\varphi\,(\sin\varphi\,d\Phi/d\varphi) + [l\,(l+1) - m^2/\sin^2\varphi]\,\Phi = 0, \\ & \qquad\qquad\qquad \text{when } s_A = s = 0,\end{aligned}$$

$$\begin{aligned}\text{(6-15)}\quad l\,(l+1) &= -p^2/\sin^2\varphi + (p+2q)\,(p+1) - q\,(q-1)\tan^2\varphi + m^2/\sin^2\varphi, \\ &= (m+2q)\,(m+1) - q\,(q-1)\tan^2\varphi, \quad \text{for } \Phi(\varphi) = \sin^p\varphi \cos^q\varphi,\end{aligned}$$

---

† The wave function Ψ (and thus R(r) also) is dimensionless, to be multiplied by a constant length for its amplitude. When we substitute $R(r) = r^b R_1(r)$, the dimension of $R_1(r)$ is r^{-b}. The same procedure applies to substitutions for $R_1(r)$.

(6-16) $\Phi(\varphi) = C_l \sin^m\varphi$, for $s_A = s = 0$, C_l is a constant, $q = 0$, and m is a half-integer or integer.

Next considering non zero values of s_A, but with $s = 0$, let us attempt a solution for $\Phi(\varphi)$ of the form of expression 6-17. The form satisfies our non ambiguity constraint $\Phi^2(\varphi) = \Phi^2(\varphi + 2\pi)$ if s_A is an imaginary half-integer or integer, *i.e.*, $s_A = n\,(-1)^{1/2}/2$ where n is an integer, and reverts to expression 6-16 when s_A is zero. Allowing s_A to be an integral multiple of $\pm\, i/2$ does not change the value of the vector potential, but permits states based on multiples of ~~σ~~/2. Inputting the form to equation 6-10 leads to expression 6-18 for $f_A(\varphi)$. Making use of the identity $(1 - \cos\varphi)/\sin\varphi = \tan(\varphi/2)$, we see that $l\,(l + 1)$ will be $m\,(m + 1)$ in the limit as φ approaches zero in equation 6-19.

(6-17) $\Phi(\varphi) = C_l \sin^m\varphi \exp(-s_A\varphi)$, where C_l is a constant, m is a real half-integer or integer, and s_A is an imaginary half-integer or integer,

(6-18)
$$\begin{aligned} f_A(\varphi) &= [1/(\Phi \sin\varphi)]\,[d/d\varphi\,(\sin\varphi\, d\Phi/d\varphi)] - m^2/\sin^2\varphi \\ &\quad + 2\,(s_A \cos\varphi)\,(1/\Phi)\, d\Phi/d\varphi + s_A\,(1/\sin\varphi - \sin\varphi) \\ &= -\,m\,(m + 1) + s_A\,(2m + 1)\,[(1 - \cos\varphi)/\sin\varphi - \sin\varphi] \\ &\quad + 2{s_A}^2\,(1 - \cos\varphi), \qquad \text{for } s = 0, \end{aligned}$$

(6-19)
$$l\,(l + 1) = -\,f_A(\varphi) + {s_A}^2 \sin^2\varphi + s_A \sin\varphi, \quad \text{when } s = 0, \text{ and } s_A \neq 0,$$
$$\lim_{\varphi\to 0} l\,(l + 1) = m\,(m + 1).$$

The sign of $\cos\varphi$ is critical to the convergence of $(1 - \cos\varphi)/\sin\varphi$ in expression 6-18, which diverges as φ approaches π. A solution of the form $\Phi(\varphi) = \sin^m\varphi \exp(s_A\varphi)$ changes the sign of the term $(1/\Phi)\, d\Phi/d\varphi$ in 6-18 and avoids the divergence at $\varphi = \pi$, but $f_A(\varphi)$ now diverges as φ approaches zero. We conclude that a single form of solution exists and that the domain for φ excludes either 0 or π. Taking the view that expression 6-17 is the elevation angle solution $\Phi(\varphi)$ for all values of s_A and s, and inserting it into equation 6-13, we obtain expression 6-20 and equation 6-21, which lead to the form of equation 6-22.

$$(6\text{-}20)\quad f(\varphi) = f_A(\varphi) + 2s\,(\sin\varphi)\,(1/\Phi)\,d\Phi/d\varphi + s\cos\varphi + s^2$$
$$= -m\,(m+1) + 2s\,m\cos\varphi + s\cos\varphi + s^2 + 2s_A{}^2\,(1-\cos\varphi)$$
$$+ s_A\,[(2m+1)\,[(1-\cos\varphi)/\sin\varphi - (2s+2m+1)\sin\varphi],$$

$$(6\text{-}21)\quad l\,(l+1) = m\,(m+1) - s^2 - s\,(2m+1)\cos\varphi - 2s_A{}^2\,(1-\cos\varphi)$$
$$+ s_A\,[(2m+2s+1)\sin\varphi - (2m+1)\,(1-\cos\varphi)/\sin\varphi]$$
$$+ (s_A\sin\varphi - s\cos\varphi)^2 + (s_A\sin\varphi - s\cos\varphi),$$

$$(6\text{-}22)\quad \lim_{\varphi\to 0} l\,(l+1) = (m-2s)\,(m+1), \qquad \text{where } Q = 0 \text{ at } \varphi = \delta = 0,$$
$$\equiv (m-s)\,(m-s+1) - s\,(s+1)$$
$$= j\,(j+1) - s\,(s+1) = J_{orb}{}^2/\sigma^2, \qquad \text{for } j = m - s.$$

As the coefficient of $1/r^2$ in the energy equation, we identify $l\,(l+1)\,\sigma^2$ evaluated at $\varphi = 0$ as $J_{orb}{}^2$, the square of specific orbital angular momentum. The angle φ is zero at γ_{oz}, where Q, ξ', and δ become zero as $\underline{\mathbf{J}}_{sum}$ aligns with $\underline{\mathbf{J}}_{oz}$. We further identify $[j\,(j+1)]^{1/2}\,\sigma\,\underline{\mathbf{k}}_o$ as $\underline{\mathbf{J}}_{oz}$, which remains constant even when the angle γ between $\underline{\mathbf{k}}_o$ and $\underline{\mathbf{J}}_{orb}$ varies in the body frame. Requiring $J_{orb}{}^2 = (m-2s)\,(m+1)\,\sigma^2$ to be an integral multiple of $\sigma^2/4$ constrains the values of s, consistent with the values of m. Since a factor of $\sin\varphi$ always appears at the lowest solution level for Φ when $m \neq 0$, an objection might be raised to the fact that our solution for Φ will be zero when φ is zero. However, it is the limit of derivatives of Φ divided by Φ that enter our computations. We also note that there exists a regular series solution for Φ for real non-integral values of m which becomes unity at $\varphi = 0$ and diverges at $\varphi = \pi$. † Such a form could be used for Φ, but we will use our indicated form, which provides results that agree with observations.

Requiring $J_{orb}{}^2$ to be an integral multiple of $\sigma^2/4$ seems to disallow values of s other than 0 and −1. However, s = 1/2 leads to $J_{orb}{}^2 = l\,(l+1)\,(2\sigma)^2$, where $l = m/2 - 1/2$. Similarly, we obtain $(m-2s)\,(m+1)\,\sigma^2 = l\,(l+1)\,\sigma^2/4$ for s = −1/4, where $l = 2m+1$. The approach for s = 1/2 also works for s = −3/2 with $l = m/2 + 1/2$, and the result for s = −1/4 also works for s = −3/4 with $l = 2m + 2$. Thus, a collection of permitted s-values exists for which

† See Hildebrand, *Advanced Calculus For Engineers*, *supra*, page 191, problem 25.

J_{orb}^2 will be an integral multiple of $\sigma^2/4$. We conclude that stable solutions for $\Phi(\varphi)$ exist throughout the field, and that body configurations wherein $\gamma \neq \gamma_{oz}$ and a small component of $\underline{\mathbf{J}}_{oy}$ is present are not precluded. But it is only when φ is zero at $\gamma = \gamma_{oz}$ that the body and inertial configurations align.

The Radial Solution

Rewriting equation 6-11 as 6-23, where $l\,(l+1)\,\sigma^2 = J_{orb}^2$ and $Q = 0$, we know from classical theory that the term in brackets multiplying R_1 is $r'^2/2$, which may become zero but is never negative. We also know that E_n/μ_p is $-Mg_o/(2a_n)$ and J_{orb}^2 is $Mg_o a_n (1-\varepsilon^2)$, where a_n is the orbit's semi-major axis and ε is its eccentricity. Making these substitutions and multiplying by r^2/σ^2, we obtain 6-24. If ε is zero, the multiplier of R_1 vanishes, and the solution is $R_1 = b_1/r + b_o$, where b_1 and b_o are appropriate constants. When ε is not zero, r oscillates in the range $a_n(1-\varepsilon) \leq r \leq a_n(1+\varepsilon)$, with $r' = 0$ at the extremes.

(6-23) $(\sigma^2/r^2)\, d/dr\,(r^2\, dR_1/dr) + 2\,[-J_{orb}^2/(2r^2) + Mg_o/r + E_n/\mu_p]\, R_1 = 0,$

(6-24) $d/dr\,(r^2\, dR_1/dr) + [Mg_o/(a_n\sigma^2)]\,[\varepsilon^2 a_n^2 - (r-a_n)^2]\, R_1 = 0,$
where $E_n = -\mu_p Mg_o/(2a_n)$ and $J_{orb}^2 = Mg_o a_n (1-\varepsilon^2)$.

Next setting $R_1 = R_2(r)/r$, we obtain equation 6-25, where $R_2(r)$ is an explicit function of r alone. If we define ω_r as the average value of the positive term $[Mg_o/(a_n\sigma^2)]^{1/2}\,[\varepsilon^2 a_n^2 - (r-a_n)^2]^{1/2}$ as r oscillates, $R_2(r)$ will exhibit oscillatory motion for r as the independent variable, equation 6-26. Expression 6-27 provides the solution for R(r). Even if we had not averaged $r'^2/2$, the function $R_2(r)$ would still be oscillatory in r, and the solution is consistent with observations for the planetary orbits.

(6-25) $d^2R_2/dr^2 + [Mg_o/(a_n\sigma^2)]\,[\varepsilon^2 a_n^2 - (r-a_n)^2]\, R_2 = 0,$ for $R_1 = R_2(r)/r,$

(6-26) $R_2(r) = b_o \cos(\omega_r r + \alpha),$ where b_o and α are constants, and
$\omega_r = [Mg_o/(a_n\sigma^2)]^{1/2}\,[\varepsilon^2 a_n^2 - (r-a_n)^2]^{1/2},$

(6-27) $R(r) = b_o\, r^{s-1} \cos(\omega_r r + \alpha),$ where $-s_A \sin\varphi + s\cos\varphi = s$ at $\varphi = 0$.

APPENDIX 7 – MOON MOTION EFFECTS ON EARTH'S ORBIT

In Chapter 7 we allowed for small frame rotations of ϕ' about the $\underline{\mathbf{k}}_o$ axis and γ' about the $\underline{\mathbf{i}}_o$ axis, in addition to ξ' about the Earth's spin axis $\underline{\mathbf{k}}$. See Figure 7.2 above. The small frame rotations are due to a solar torque on the Earth's body, and to the Moon's motion. We shall now derive the latter.

Torque On The Earth's Body Due To The Moon's Motion

The small frame rotations of ϕ'_m and α'_m we derived in Appendix 3 for the Moon's orbit vector $\mu_m\underline{\mathbf{J}}_m$ are produced by a solar torque on the Moon's orbit. They cause $\underline{\mathbf{J}}_m$ to precess and nutate about the Earth's orbit vector $\underline{\mathbf{J}}_o$ in an Earth-based reference frame depicted in Figure A7.1. [†] The figure reflects a steady precession and nearly constant inclination of $\underline{\mathbf{J}}_m$ relative to $\underline{\mathbf{J}}_o$ in the $\underline{\mathbf{i}}_m\underline{\mathbf{j}}_m\underline{\mathbf{k}}_m$ frame we have selected. The $\underline{\mathbf{j}}_m$ axis aligns with $\underline{\mathbf{j}}_o$ when the spin axis $\underline{\mathbf{s}}$ lies in the $\underline{\mathbf{j}}_o\underline{\mathbf{k}}_o$ plane. Rotation about $\underline{\mathbf{j}}_o$ by the angle ι aligns $\underline{\mathbf{k}}_m$ with $\underline{\mathbf{J}}_o$ to produce our chosen frame. To first order, we found the precession frequency to be $\phi'_m \cong 2\pi$ radians per 18.6 years, and the inclination α_m of $\underline{\mathbf{J}}_m$ to $\underline{\mathbf{J}}_o$ to oscillate very slightly about a median value of 5.1 degrees over a six month period due to the solar torque on the lunar orbit slaved to the Earth.

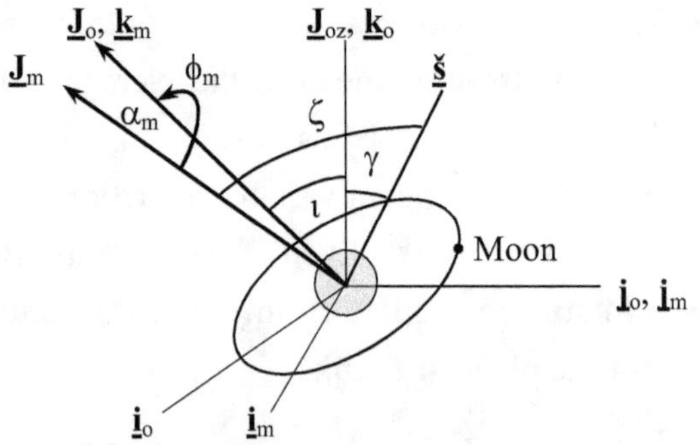

Figure A7.1. The Moon's Orbit In The Earth's Body Frame

[†] Since the center of mass for the Earth and Moon is approximately 4,728 kilometers from the Earth's center, we have a slight geometry error.

However, the Moon's motion, in turn, places a torque on the Earth's orbit which depends on the orientation of the Moon's orbit vector $\underline{\mathbf{J}}_m$ at any time, and on the location of the Moon itself, as specified by its orbit frequency w_m. The quadrupole solution for Laplace's equation models a scalar potential V_{b2} due to the lunar motion, expression 7-1, which is based on the angle ζ between $\underline{\mathbf{J}}_m$ and the spin $\underline{\check{\mathbf{s}}}$ depicted in Figure A7.1. Using the Earth's mass μ to substitute $\mu_m \cong \mu/81.30$, † replacing $\mu g_o/r_m^3$ by w_m^2, and setting $w_m \cong 2\pi$ radians per 27.32 days $\cong 13.37\ W_e$, we obtain expression 7-1a. V_{b2} is not an explicit function of W_e, we are simply using it to compute V_{b2}.

(7-1) $V_{b2} = -(I_s - I_a)(\mu_m g_o/r_m^3)(3\cos^2\beta_2 - 1)/2,$

where r_m is the Moon's orbit radius, μ_m is its mass, $\cos\beta_2 = \sin\zeta\,\sin(w_m t - \phi'_m t)$ for its orbit frequency of w_m and precession frequency of ϕ'_m, ζ is the angle between $\underline{\mathbf{J}}_m$ and the Earth's spin direction $\underline{\check{\mathbf{s}}}$, I_s is the Earth's moment about $\underline{\check{\mathbf{s}}}$, and I_a is the Earth's moment about any equatorial plane axis,

(7-1a) $V_{b2} = -(I_s - I_a)(w_m^2/81.30)[3\sin^2\zeta\,\sin^2(w_m t - \phi'_m t) - 1]/2,$

$= -2.20\ W_e^2 (I_s - I_a)[3\sin^2\zeta\,\sin^2(w_m t - \phi'_m t) - 1)]/2,$

where $W_e = 2\pi$ radians per year,

(7-2) $I_s\underline{\check{\mathbf{s}}} = I_s\check{s}_o(-\sin\iota\cos\gamma\,\underline{\mathbf{i}}_m + \sin\gamma\,\underline{\mathbf{j}}_m + \cos\iota\cos\gamma\,\underline{\mathbf{k}}_m),$

(7-3) $\underline{\mathbf{J}}_m = J_m(\sin\alpha_m\cos\phi'_m t\,\underline{\mathbf{i}}_m + \sin\alpha_m\sin\phi'_m t\,\underline{\mathbf{j}}_m + \cos\alpha_m\,\underline{\mathbf{k}}_m),$

where $\alpha_m \cong 5.1$ degrees, and $\phi'_m \cong 2\pi$ radians per 18.6 years,

(7-4) $\cos\zeta = \sin\alpha_m(-\sin\iota\cos\gamma\cos\phi'_m t + \sin\gamma\sin\phi'_m t) + \cos\iota\cos\gamma\cos\alpha_m.$

Using expressions 7-2 and 7-3 for $I_s\underline{\check{\mathbf{s}}}$ and $\underline{\mathbf{J}}_m$ in the frame of Figure A7.1, where t is an elapsed time since ζ was measured, equation 7-4 specifies ζ for the dot product of $\underline{\mathbf{J}}_m$ and $\underline{\check{\mathbf{s}}}$. If we square $\cos\zeta$ and average over the lunar

† See, Selby, S. (ed.), *CRC Standard Mathematical Tables*, 15th Ed., The Chemical Rubber Co., Cleveland, OH (1967), pp.479-481, for the Moon's parameters.

precession, all terms having single factors of $\cos\phi'_m t$ or $\sin\phi'_m t$ or $\cos\phi'_m t$ $\sin\phi'_m t$ average to zero, leaving expressions 7-4a and 7-5 for $\cos^2\zeta$ and $\sin\zeta$ $\cos\zeta\ \partial\zeta/\partial\gamma$ as mean values. Although the terms vary slightly over an 18.6 year period, we will use the mean to compute $\partial V_{b2}/\partial\gamma$. Inputting $\alpha_m \cong 5.1$ degrees and $\iota \cong 11.31$ degrees, equation 7-6 provides the partial derivative of V_{b2} with respect to γ. When we average over the orbit, a factor of 1/2 replaces $\sin^2(w_m t - \phi_m)$ for $\phi_m = \phi'_m t$, and this is the form used in Chapter 7 to compute the Moon's contribution to the precession of the Earth's equinoxes.

(7-4a) $\cos^2\zeta \cong (\sin^2\alpha_m)/2 + 0.9881 \cos^2\iota \cos^2\gamma$,

where $(3\cos^2\alpha_m - 1)/2 = 0.9881$,

(7-5) $\sin\zeta \cos\zeta\ \partial\zeta/\partial\gamma \cong 0.9881 \cos^2\iota \sin\gamma \cos\gamma$, mean value,

(7-6) $\partial V_{b2}/\partial\gamma = -6.271\ W_e^2\ (I_s - I_a) \sin\gamma \cos\gamma \sin^2(w_m t - \phi'_m t)$.

APPENDIX 8 –SMALL FRAME ROTATIONS FOR MARS

The mean orbit radius for Mars is $a_n \cong 2.259 \times 10^8$ kilometers, which is about 1.52 times the Earth value, and its present orbital angular frequency is $W_m \cong 2\pi$ radians per 1.881 Earth years. Using a mean planet radius of h_a = 3,357 kilometers and a J_2 value of 1958.6×10^{-6} determined from Viking orbiter data in 1995, † we obtain $\epsilon_m = [(I_s - I_a)/I_z]^{1/2} = 0.6577 \times 10^{-6}$.

Our wave theory provides Martian parameters of $\mu J_{morb} = \mu(32)^{1/2}\,\varpi/2$ and $\mu J_{moz} = \mu(35)^{1/2}\,\varpi/2 \cong 1.04583\, I_{zm} W_{mo}$, with γ_{moz} = arc-cos$(32/35)^{1/2} \cong$ 17.024 degrees. Equation 8-1 thus specifies W as a function of γ for Mars, where $W = W_{mo}$ and $\xi'_{mo} = 0.04815\, W_{mo}$ occur at $\gamma_{mo} \cong 17.852$ degrees. The angle ι_m for Mars is 9.46 degrees, and its observed obliquity of $\nu_m = 25.20$ degrees specifies the present value of γ for Mars at $\gamma_m = 23.465$ degrees. Using γ_m to calculate $W_m = 0.81064\, W_{mo}$, we find that $W_{mo} = 2\pi$ radians per 1.525 Earth years. However, we can use the presently observed value of $W_m = 2\pi$ radians per 1.881 Earth years to calculate W in expression 8-1.

(8-1) $W = W_{mo}(1.04583 - \cos\gamma)/\sin^2\gamma$, where $\gamma_{mo} = 17.85$ degrees,
$W = 1.2336\, W_m(1.04583 - \cos\gamma)/\sin^2\gamma$, using $\gamma_m = 23.465$ degrees.

Based on Pathfinder and Viking data, JPL analysts have calculated the constant precession rate for Mars at $\phi'_{mobs} = -7.576$ arc-seconds per year $\cong -16.719\, \epsilon_m W_m$, and have estimated I_s to be $0.366\, \mu h_a^2$ without considering any effects of γ'' or changes in W for the solar torque equation. †† If we set γ'' to zero at γ_m in the body equation for Mars with $\check{s}_{mo} = 2\pi \times 365.25/1.029$ radians per year, and then set $W_c = W_m$, we obtain equation 8-2 when we

---

† See, Konopliv, A., and Sjogren, W., JPL Pub. 95-3, California Inst. Tech. (1995); see also, Smith, D., *et al.*, *J. Geophys. Res.* 98, 20871 (1995). We note that a measurement of J_2 provides the difference between I_s and I_a, and not the value of either.

†† See, Folkner, W., Yoder, C., Yuan, D., Standish, E., and Preston, R., "Interior Structure and Seasonal Mass Redistribution of Mars from Radio Tracking of Mars Pathfinder", *Science*, 5 Dec 1997, vol. 278, no. 5344, pp. 1749-1752. The analysis states in the text that "the estimated obliquity rate is consistent with zero", but at note 15 attributes –100 milli-arc-seconds of precession to "un-modeled nutations".

express the term $I_s š_{mo} \phi'_{mobs}/I_z$ as $3.748\, C_m \in_m^2 W_m^2$ for $I_s = C_m\, \mu h_a^2$. Such an approach assumes that C_m may be computed at $\gamma_m = \gamma_{mc}$. Our body equations would then show C_m to be about 0.357 rather than 0.366, primarily due to multiplying $\cos\gamma_m$ by $\cos^2\iota$ instead of $\cos\iota$ in equation 8-2.

$$(8\text{-}2) \quad I_s š_{mo}\, \phi'_{mobs}/I_z = 3.748\, C_m \in_m^2 W_m^2 = 1.5 \in_m^2 W_m^2 \cos^2\iota \cos\gamma_{mc},$$
$$\text{where } \in_m = 0.6577 \times 10^{-6}.$$

However, inputting ϕ'_{mobs} alone to the Martian equation for precession does not determine I_s. Instead, the equation is based on γ_{mc}, where γ'^2 is at its maximum, which is most likely not γ_m. We could instead use hypothetical values of γ_{mc} in equation 8-3, with equation 8-1 specifying W_{mc}/W_m, to estimate C_m. As examples, we find that C_m would be 0.346 if $\gamma_{mc} = 24.0$ degrees, and 0.240 if $\gamma_{mc} = 33.3$ degrees. † Substituting $\gamma'^2/4$ for E_γ/I_z and again ignoring $I_a(\phi'^2 \sin^2\gamma)/2$, the body energy is given by equation 8-4 for Mars.

$$(8\text{-}3) \quad C_m = 0.4002\, (W_c/W_m)^2 \cos^2\iota \cos\gamma_{mc},$$

$$(8\text{-}4) \quad \gamma'^2/(4 \in_m^2 W_m^2) = E_b/(I_z \in_m^2 W_m^2) + 3.7481\, C_m \cos\gamma - 0.5\, (W/W_m)^2.$$

If the present obliquity rate is $\nu'_m = -0.100$ arcseconds per year, as JPL analysts seem to have modeled, γ'_m would be about -0.108 arc-seconds per year, expressible as $-0.2383 \in_m W_m$. Further assuming that $C_m = 0.366$, †† we obtain $E_b/I_z = -0.7442 \in_m^2 W_m^2$ when we input $\gamma'^2_m/4 = 0.0142 \in_m^2 W_m^2$ to equation 8-4. The extremes of γ will then be $\gamma_{m\text{-}min} \cong 22.7$ degrees and $\gamma_{m\text{-}max} \cong 36.7$ degrees, with obliquity extremes of $\nu_{m\text{-}min} \cong 24.5$ degrees and $\nu_{m\text{-}max} \cong$

---

† The terms $I_z(W + \xi' \cos\gamma)\, \phi'$ and $I_s(š + \xi')^2/2$ in the expression of E are oppositely signed and orders of magnitude greater than other body terms. For the Earth we find that their sum is nearly zero. If the same condition should hold for Mars, it would require $C_m \cong 0.240$ and indicate the lower limit of a dense solid core for a spherical body with a mantle of approximate bulk density 2.3 grams per cubic centimeter and overall density similar to the Earth's.

†† Bills, B., *Geophys. Res. Lett.*, 16, 385 (1989), estimates I_s at $0.345\, \mu h_a^2$. See also, Reasenberg, R, *J. Geophys. Res.*, 82, 369 (1977), estimating I_s at $0.365\, \mu h_a^2$.

37.7 degrees, expression 8-5. A rough time of passage calculation yields a period of about a million years.

(8-5) 24.5 degrees $\leq \nu \leq$ 37.7 degrees, for Mars,
if $\nu'_m = -0.100$ arc-seconds per year,
and $C_m = 0.366$.

Astronomers have estimated the obliquity for Mars to range from 15 to 35 degrees over a 124,000 year cycle with excursions from 0 to 60 degrees over tens of millions of years, as compared with the JPL data estimate for ν' of roughly 3 degrees per 100,000 years. Astronomers' base estimate is more than 10 times the JPL estimate, and the two cannot be reconciled. †

The nutation results are much more sensitive to the value of γ'_m than to C_m, and the lower obliquity limits are achieved only if $|\gamma'_m|$ exceeds 0.100 arc-seconds per year. For example, setting $C_m = 0.345$ for $\gamma_c \cong 24.1$ degrees with $\gamma'_m = -0.653$ arc-seconds per year, which is about 17 percent greater than the Earth's value, we obtain an obliquity range from 15 to 63 degrees over a period of about a million years. We have no rationale for choosing among the possible values of C_m on the basis of currently available data; †† however, the nutation of Mars' spin axis will not be chaotic for any reasonable value of C_m so long as the value of γ'_m is in the range being considered.

The essential point is that body motions of spinning planets are coupled to their orbital motions, and cannot be treated as free spinors subjected to various torques. Previous estimates of body moments based on precession calculations are correct only when the nutation rate is near its maximum.

---

† See, Moomaw, B., "The Obliquity of Mars", http://www.spacedaily.com/news /mars. html, 30 Jun 00; see also, Beatty, et al., *The New Solar System*, *supra*, Fig. 20, at p. 189, author Jakosky, B.M., estimating a base obliquity range of 11 to 27 degrees with excursions of 2 to 35 degrees. See further, Ward, W., "Large-Scale Variations In The Obliquity Of Mars", *Science*, 20 Jul 73, vol. 181, no. 4096, pp. 260-262, estimating the Martian obliquity to range from 14.9 to 35.5 degrees.

†† The rover *Spirit* became inoperable in 2010 and unable to provide the data. See, http://www.science.nasa.gov/sciencenews/scienceatnasa/2010/ 4feb_martiancore/.

www.ingramcontent.com/pod-product-compliance
Lightning Source LLC
Chambersburg PA
CBHW051654170526
45167CB00001B/462
* 9 7 8 1 4 9 5 4 4 1 6 4 6 *